AF510633

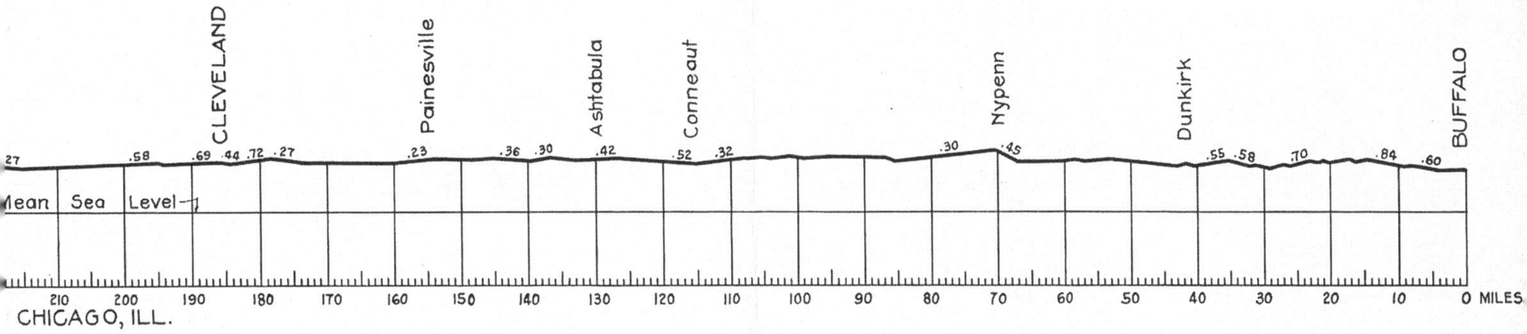

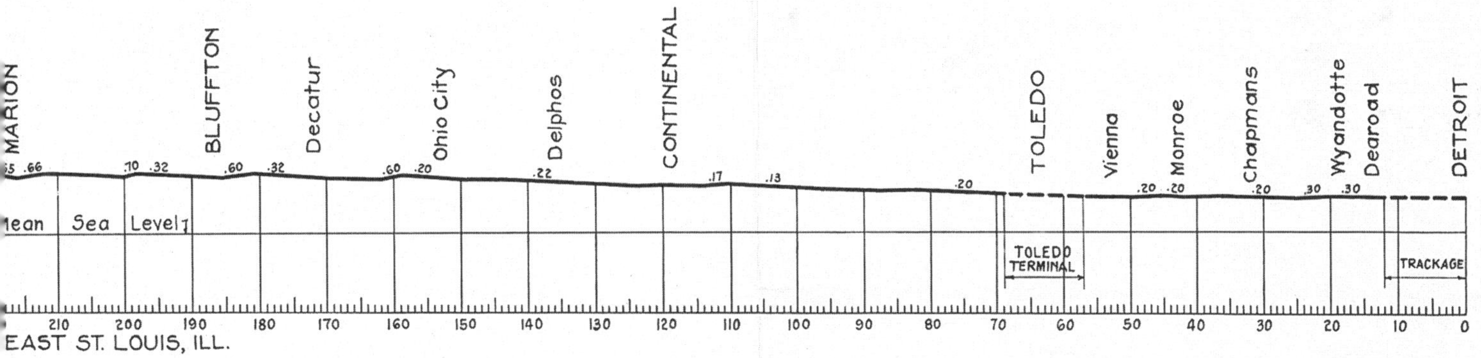

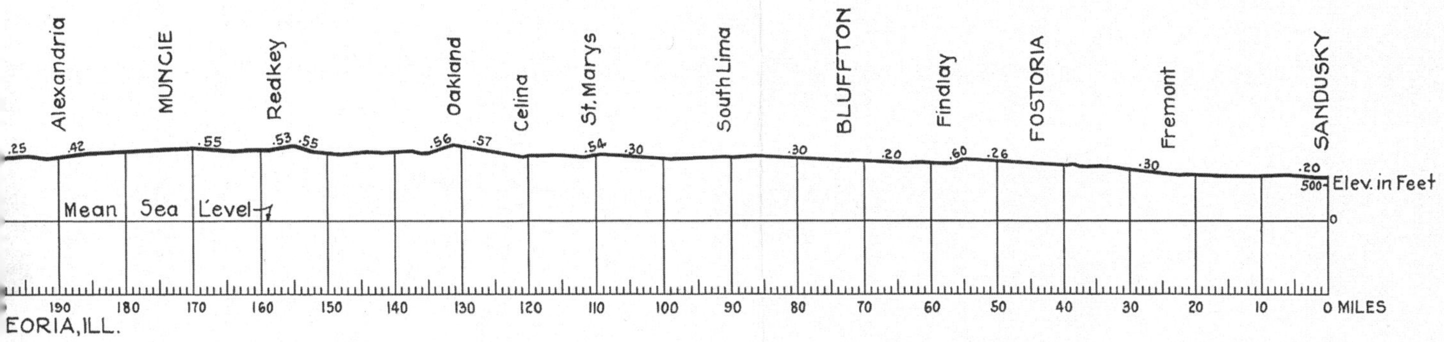

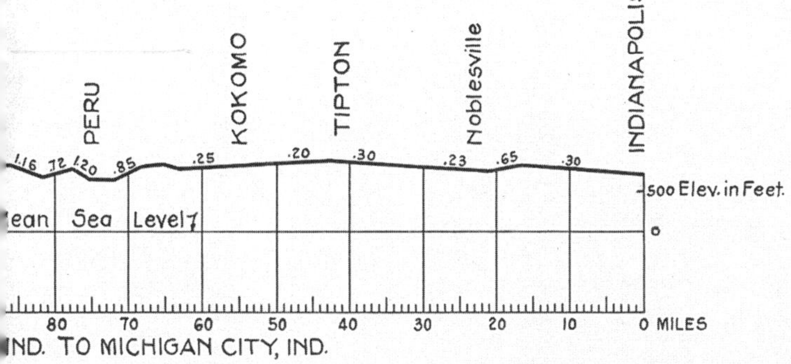

Before the INTERSTATE COMMERCE COMMISSION, Finance Docket № 7460.

THE CHESAPEAKE AND OHIO RAILWAY CO
Chief Engineer's Office ~ Richmond, Va

CONDENSED GRADE PROFILE
OF THE
NEW YORK, CHICAGO AND ST. LOUIS R.R.

DATE: DEC. 31, 1928. DRAWN BY: E.W.C. CHECKED:	DRAWING № 8810

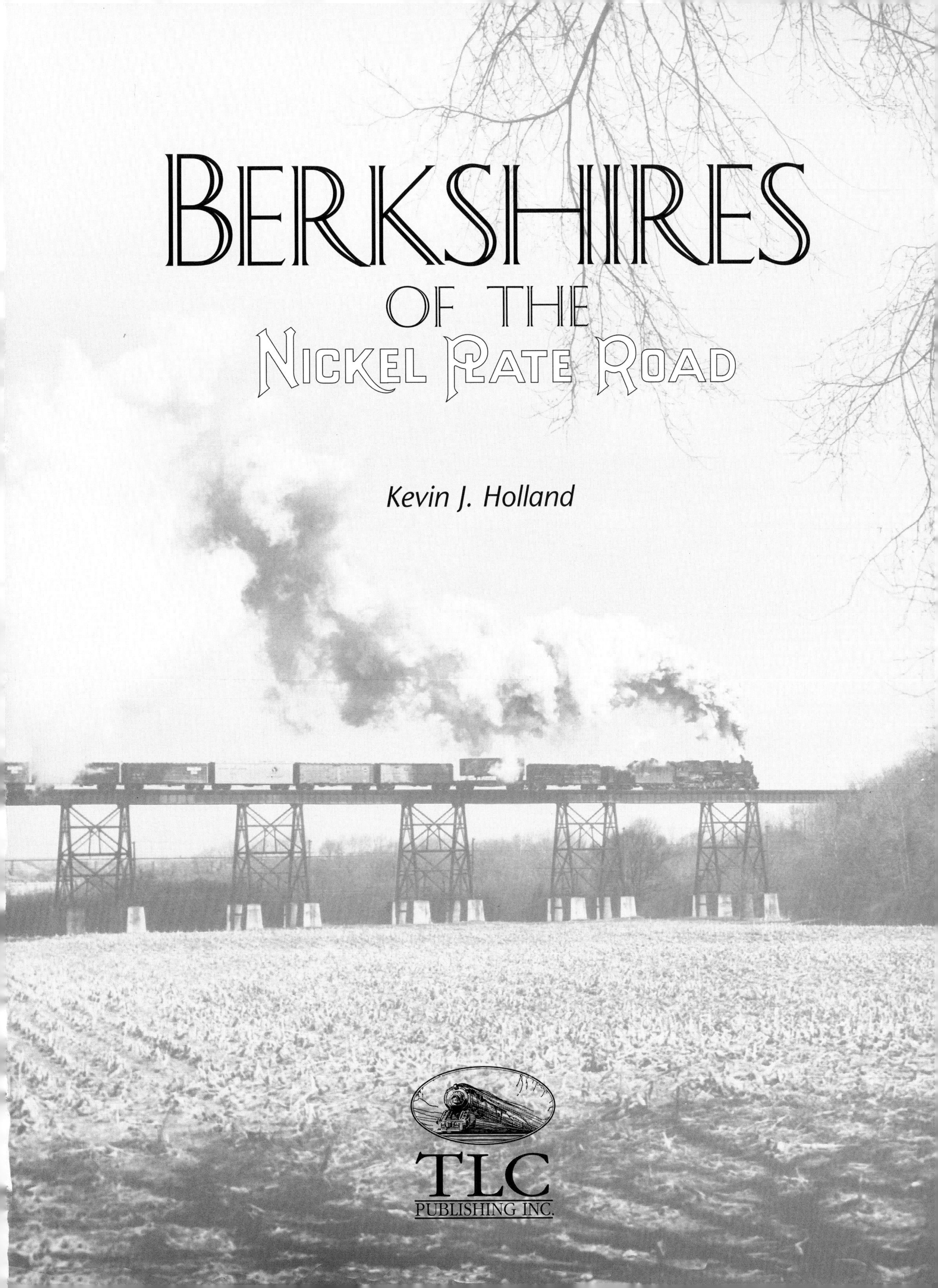

BERKSHIRES OF THE NICKEL PLATE ROAD

Kevin J. Holland

TLC
PUBLISHING INC.

1998
TLC Publishing, Inc.
1387 Winding Creek Lane
Lynchburg, Virginia, 24503-3776

© Copyright 1998 TLC Publishing, Inc.
All Rights Reserved. No part of this work may be reproduced without the written permission of the publisher except for brief quotations used in reviews or other "fair use."

International Standard Book Number 1-883089-39-5
Library of Congress Catalog Card Number 98-61006

Design, Layout, Type, and Image Assembly by
Kevin J. Holland
Type&Design Associates
Burlington, Ontario

Color Image Assembly, Separations, and Printing by
Walsworth Publishing Company
Marceline, Missouri, 64658

Produced on the MacOS™

Paintings on Front Cover and pages 118-119
© Copyright Ted Rose 1994 and 1992. Reproduced with permission.

Other Nickel Plate Road Titles From TLC Publishing:

Nickel Plate Road Passenger Service – The Postwar Years
by Kevin J. Holland

Nickel Plate Road Diesel Locomotives
by Kevin J. Holland

Nickel Plate Road Publicity Photos – Volumes 1 and 2
by John B. Corns

1938 NKP Official Passenger Car Diagrams (reprint)

1950 NKP Official Passenger Car Diagrams (reprint)

1962 NKP Official Freight Equipment Diagrams (reprint)

Cover –
In *"Taking The Main,"* Ted Rose presents a Nickel Plate scene repeated countless times between 1934 and 1958. In this instance, the S-2 Berkshires meeting at Bellevue, Ohio, are 759 and 765, two noteworthy Nickel Plate ambassadors. Through the foresight, hard work, dedication, and financial support of hundreds of people, these two Berkshires have perpetuated the legacy of an important American railroad and a significant group of locomotives.

Title Page –
A Nickel Plate S-3 has livestock, perishables, and piggyback behind its tender as it rushes one of the Road's eastbound manifests across Conneaut Creek viaduct in the spring of 1957.
(Jim Shaughnessy)

Facing Page –
Alco-built S class Berkshire 711 lifts an eastbound extra out of Bellevue yard in July 1956.
(Don Wood)

to Russell and Graeme

"Thunder is good, thunder is impressive,
but it is lightning that does the work."

Mark Twain, 1908.

Contents

John A. Rehor photographs his favorite subject at Arcadia, Ohio. *(Herbert H. Harwood, Jr.; John B. Corns Collection)*

Acknowledgments

This book completes a trilogy examining three important aspects of the postwar operations of one of North America's genuinely outstanding railroads. One of the early plans for this volume was to include a photograph of every Nickel Plate Road and Wheeling & Lake Erie 2-8-4 ... a "Berkshire Mugbook" of sorts. Like many "early plans," this one proved too limiting and was discarded. The result would have been on-line scenes at many of the same locations and repetitive roster views – not necessarily a bad thing, but at odds with the goal of presenting as diverse a pictorial record of these locomotives as possible within the pages available, while at the same time documenting their development and evolution.

Simply put, this book would not have been possible were it not for the insight, expertise, and historical resources provided by good friend John B. Corns. Aside from his being a wealth of information on the Wheeling & Lake Erie, John located and eagerly shared original NKP records and correspondence. This paperwork ranged from board meeting minutes to AFE forms and shop schedules, and shed much-needed light on some nagging questions while unequivocally putting others to rest. About half of the photos on the pages that follow are from John's collection – most of these are rare builder's photos and official Nickel Plate views printed from the original negatives. John is also an avid steam hardware historian, and his passion is reflected in the builder's plate images that open several chapters (if you're curious about a particular plate's identity, just check the builder's number against each chapter's roster).

The credit for conceiving this project goes to publisher Thomas W. Dixon, Jr., who also provided a great deal of resource material relating to the development of the Super-Power concept, Lima's A-1, and the work of the AMC on its initial C&O and Nickel Plate power. I'm grateful, once again, for Tom's confidence, support, and enthusiasm, as well as for the creative freedom he gave me.

The city of Lima, Ohio, figures largely in the story of Super-Power and the Nickel Plate Berkshires. John H. Keller and Charles Bates of that city's Allen County Historical Society not only placed their archives and personal expertise at my disposal yet again, but indulged me as I poked, prodded, and photographed their pride and joy, S-3 number 779.

Jim Shaughnessy and Don Wood graciously honored my requests to include some of the results of their independent NKP sojourns.

Ted Rose's evocative watercolors are without peer, and I'm honored to include two of his Nickel Plate paintings in this book.

Retired NKP engineman Don Daily recounted his experiences with various members of the S-classes, and provided NKP paperwork that confirmed details of the Alco-built S-class Berkshire's very earliest days.

To all those who contributed to this project with critiques, ideas, information, photographs, and suggestions for further avenues of research, thank you:

Allen County Historical Society, Lima, Ohio; Howard W. Ameling; Charles Bates; Robert Bourquin II, The Timken Company, Canton, Ohio; Bill Caloroso; Chesapeake & Ohio Historical Society; Conneaut Railroad Historical Museum, Conneaut, Ohio; Joe Collias; John B. Corns; Don Daily; Thomas W. Dixon, Jr.; Willard A. Harvey, Jr.; Eugene L. Huddleston; Kevin P. Keefe, Trains Magazine; John H. Keller; Stephen L. King; Tony Koester; Jim Kreider; Mad River & NKP Railroad Museum, Bellevue, Ohio; Dr. Louis A. Marre; Doyle L. McCormack; Kenneth L. Miller; Niagara Frontier Chapter, NRHS; Nickel Plate Road Historical & Technical Society; Al Paterson; J.R. Quinn; William A. Raia; Ted Rose; Jim Shaughnessy; Peter K. Shepherd; Keith Sirman; Allan Stanley; James Van Brocklin; Margaret T. Whittington; Jay Williams, Jr.; Don Wood; Craig A. Woodworth.

BERKSH

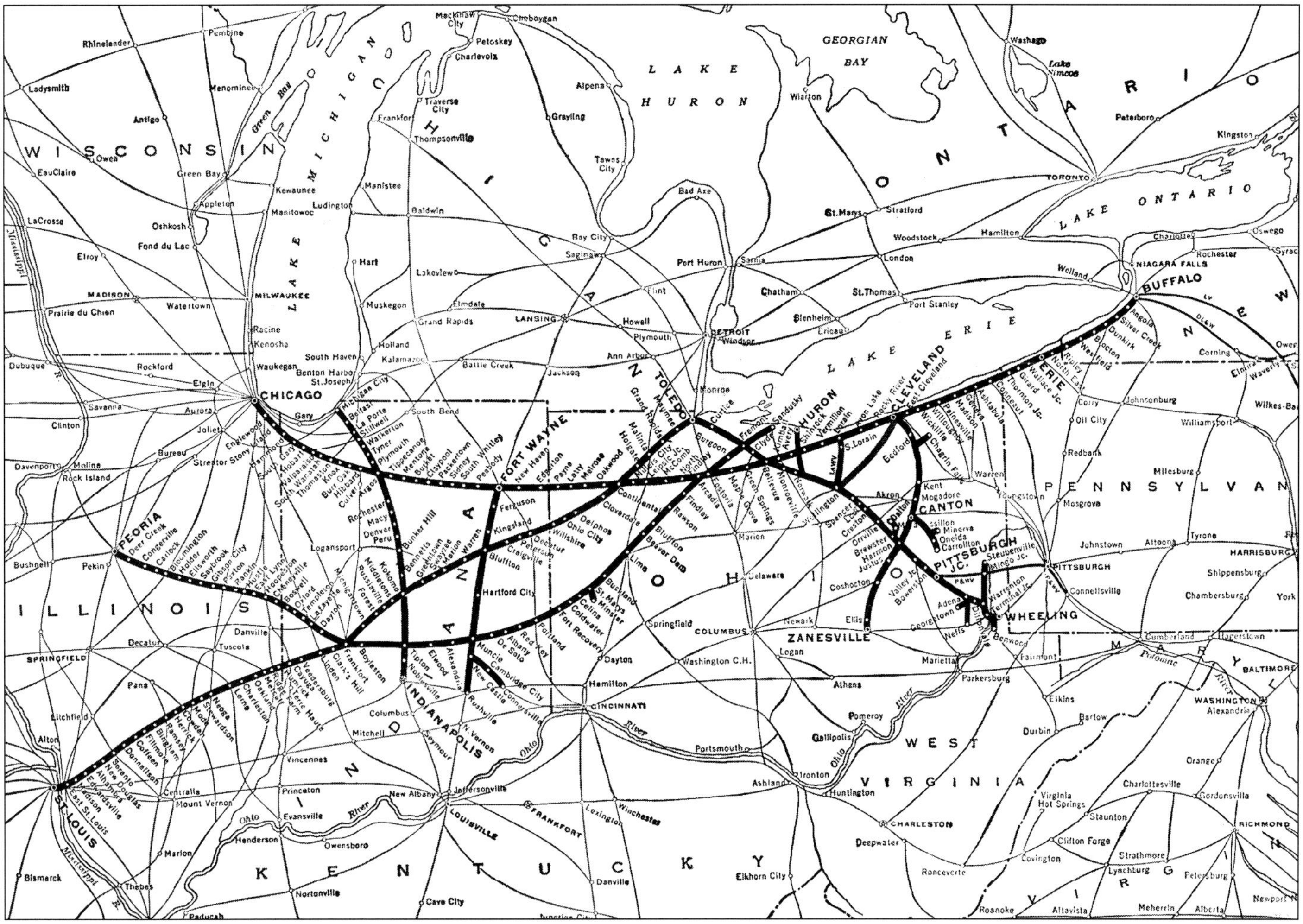

RE COUNTRY

Preface

They were, according to the late John Rehor, "the engines that saved a railroad." Few North American steam locomotive designs more thoroughly embodied the spirit and character of their owner – or were so widely respected, particularly by the competition – than the Berkshires of the New York, Chicago & St. Louis Railroad. In the capable hands of a Nickel Plate crew, these legendary locomotives carried the responsibility for the Road's post-depression solvency squarely on their shoulders. They did not disappoint, and made "Nickel Plate Road High Speed Service" the envy of the industry. To shareholders and fans alike, the "Berks" were the underlying reason for much of the NKP's substantial appeal. Without these locomotives it would have been a very different railroad.

Nicknamed by an Ohio newspaper editor in honor of its high construction standards, the Nickel Plate Road opened for business on October 23, 1882, and immediately tried to capture a share of the Buffalo-Chicago business moving on the rails of parallel competitor Lake Shore & Michigan Southern. The Nickel Plate had been built as a thorn in the LS&MS' side, and quickly proved successful. So successful, in fact, that it acquired new owners barely two months after running its first revenue train – the Vanderbilts bought the new road to protect the interests of their LS&MS, in an era before such transactions were frowned upon. Thus began a 33-year period of corporate neglect that set the stage for one of the cruelest ironies in American railroading. While the LS&MS (and, later New York Central) was developed into a fast, first class multi-track carrier with modern equipment and physical plant, the Nickel Plate Road remained a single-tracked stepchild powered by antiquated locomotives. Government caught up to the NYC in 1915, when the road was ordered to relinquish control of either the LS&MS, NKP, or Michigan Central. The Nickel Plate was sold in 1916 to Cleveland's Van Sweringen brothers, Oris P. and Mantis J., ostensibly to secure a segment of right-of-way to connect their Shaker Heights suburban transit project with downtown Cleveland. In the course of the next two decades, through their Alleghany Corporation, the brothers built this first rail holding into an empire that included control of the Chesapeake & Ohio, Wheeling & Lake Erie, Pere Marquette, Erie, Missouri Pacific, Rio Grande, Texas & Pacific, and Chicago & Eastern Illinois, as well as Cleveland's Union Terminal development. The brothers knew their limitations, though, and entrusted the management of their railroads to a pool of talented professionals, chief among them John J. Bernet (whom they had hired away from NYC in 1916). Bernet's vision took the legacy of his former employer's control – a neglected physical plant and obsolete motive power – and systematically transformed the Nickel Plate into a lean, aggressive property, administered by General Manager A. R. "Gus" Ayers. The NKP still had to fight for its traffic – which, after 1923, included business from St. Louis and Peoria over the merged Clover Leaf and Lake Erie & Western lines – but it could no longer be ignored. The Road earned a solid and enduring reputation for service and speed. Service was the byword for thousands of Nickel Platers, from the Chairman to the newest hire. Speed was reflected in the Road's pursuit of motive power excellence, which reached its highest evolution and execution with the Berkshires.

The Nickel Plate, to be sure, operated other classes of steam power, and a number of competing roads also operated locomotives sharing the Berkshires' 2-8-4 wheel arrangement. Yet nowhere else was the essence of owner and machine merged so seamlessly and productively as it was between 1934 and 1958 in Berkshire Country – the Midwestern swath that came to know the Nickel Plate's High Speed Freight Service first-hand.

As steam propulsion on American railroads reached its zenith in the years before and dur-

ing World War II, a company's premier locomotives quite often became an integral element of the line's public identity. Southern Pacific's Daylights were pure Hollywood, and shared the public spotlight with SP's unique Cab-Forwards. New York Central – the Water Level Route – had its fluvially named Hudsons, Mohawks, and Niagaras; Union Pacific, its Challengers and Big Boys. Chesapeake & Ohio was linked in the public eye to ponderous Alleghenies and dignified Greenbriers; Pennsy to its elegant but powerful K-4 Pacifics and brutish M-1 Mountains. That the apogee of 2-8-4 design – Nickel Plate's 700s – would forever be known by a name recalling a staunch competitor is a curious quirk of fate. Coming up with a "new" name was as much an unecessary frill for the pragmatic Nickel Plate as was the semi-streamlined shrouding proposed for the first 15 NKP Berks. For the Nickel Plate Road – the railroad that elevated the 2-8-4 to its greatest success – there was a degree of irony in the name by which that wheel arrangement had become generally known. The first 2-8-4s went into service for New York Central subsidiary Boston & Albany, where Lima demonstrator A-1 had shown its prowess hauling freight up and over the mountains of western Massachusetts – the Berkshire Hills. Thus were the first Berkshires christened. New York Central certainly drew no comfort from this nomenclative victory, as the Nickel Plate's 2-8-4s regularly upstaged Central's best efforts, whether pulled by steam or diesel. And it wasn't that NYC's best steam locomotives were incapable, either – nor those of the Wabash, Pennsy, or Erie, for that matter. The Nickel Plate and its Berkshires just made them look that way.

Breaking in on train 35 near Kingsville, Ohio, September 1, 1957.
(John A. Rehor; John B. Corns Collection)

For the NKP, that was just part of a day's work – the triumvirate of bridge traffic, speed, and Berkshires. Each was essential to the Road's survival during and after the Great Depression – remove or impede one element, and the other two were moot. The formula was honed so perfectly by crews and successive Nickel Plate managements that the Berkshires lasted until the summer of 1958, in the frontline service for which they had been built. Even then, they were withdrawn with the expectation that some would return to assist with an autumn traffic surge that failed to materialize. The Nickel Plate was fully, and finally, dieselized by the summer of 1959 – and Berkshire Country would never be quite the same. The Nickel Plate itself disappeared into Norfolk & Western in October 1964, leaving a legacy of service and innovation rarely matched even today. Thankfully, the Road also left us a few Berkshires. In every sense, they were the right locomotives – conceived, designed, and built at the right time – for the right railroad.

As this is written, fifty years to the day after the last NKP Berkshires were ordered from Lima and forty years to the day after the last 700 made a revenue run for the Nickel Plate, large black locomotives still follow the shadows of the Berkshires between Buffalo and Chicago – faithful and worthy heirs in objective if not in spirit. If you are fortunate enough to have known these locomotives first-hand, the pages that follow offer a chance to return trackside. Whether circumstances caused you to admire the Berkshires from afar or, like so many over the past thirty years, you came to know the survivors in excursion service and museums, then consider these pages a celebration of the locomotives that made the Nickel Plate such a successful – and memorable – railroad.

Kevin J. Holland
July 2, 1998

Chapter 1 Woodard, Bernet

While demonstrator locomotives are often considered a product of the diesel age, few had a broader impact than Lima's A-1. Outshopped in 1925 – and shown here lettered for Boston & Albany – the first 2-8-4 triggered a revolution in railroad operations surpassed only by Electro-Motive's FT fourteen years later. *(Lima Locomotive Works; Allen County Historical Society Collection)*

Railroading's "drag era" had plateaued by the 1920s with the widespread use of Consolidation (2-8-0 wheel arrangement) and – for more mountainous territory – Decapod (2-10-0) locomotives, at least on those roads which had not adopted articulated steam power. Tonnage moved, but relatively slowly. During the drag era, train speeds of about 35 miles per hour over level ground were all that could be expected of most freight locomotives, whose strength lay in their ability to start heavy consists. The penalty for this high tractive power (relative to the locomotives' weight) was steeply diminished horsepower once these drag engines got their trains rolling.

Increases in locomotive speed and power were needed, particularly for railroads like the New York, Chicago & St. Louis, which by

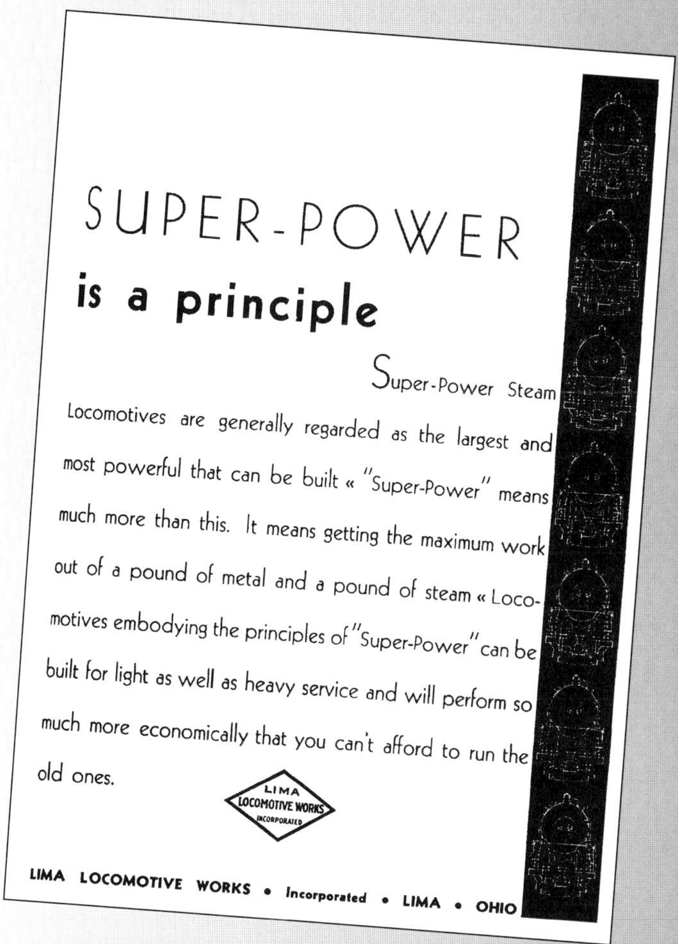

AND THE AMC

the early 1930s offered a classic study of a bridge line in very competitive territory literally being starved for lack of locomotives that could expedite fast, frequent trains of moderate tonnage. With approximately two-thirds of its traffic received from connecting lines, and most of that in turn forwarded to still more connections, the Nickel Plate was forced to evolve without the financial cushions of strong originating or terminating traffic. It had to survive on speed and service.

The Mikado (2-8-2) was a design evolution that used a larger firebox (supported by a two-wheel trailing truck) to provide more steam and hence greater sustained horsepower. The 2-8-2 was pioneered by Baldwin in the late 1890s with an order of narrow gauge engines exported to Japan – the Asian nickname stuck even after domestic examples had been built, although in time the wheel arrangement was referred to just as often as the Mike. The 2-8-2 continued to evolve – in part through the development of efficiencies like the mechanical stoker – and it reached wide acceptance in the USRA heavy and light Mikado designs of World War I.

The commercial construction of steam locomotives in the United States during the twentieth century fell, for the most part, to the "Big Three" builders: Baldwin, of Philadelphia (Eddystone); the American Locomotive Company, or Alco, based in Schenectady, New York, but with affiliated "works" in several cities; and the Lima Locomotive Works of Lima, Ohio. Some large railroads, notably Norfolk & Western and Pennsylvania, also built significant numbers of steam locomotives in their own shops.

Lima was the smallest builder of the Big Three, but made up for any handicap in that regard by being an innovator, driven by the desire to create products of superior performance and quality. While the phrase "pushing the envelope" had not yet been coined, it would certainly have applied to Lima. Much like the Nickel Plate, it was an underdog in its market, forced to be that much more resourceful and aggressive than its larger competitors. Tracing its origins back to 1869 as a manufacturer of agricultural implements and stationary boilers, Lima gained wide recognition after 1880 as the builder of Ephraim Shay's patented geared locomotives.

The construction of conventional rod-driven steam locomotives accounted for more and more business as the company grew during the first decades of the twentieth century, and, not content with the status quo of reciprocating steam design, the technicians and managers at Lima sought methods to improve the very essence of locomotive performance. In the process, they transformed railroading in a manner not to be seen again until the 1939 debut of Electro-Motive's FT.

W. E. Woodard.

(Allen County Historical Society Collection)

William E. Woodard was perhaps the greatest personification of Lima's well-earned status as an innovator. Born in Utica, New York, in 1873, Woodard graduated from Cornell University in 1896 with a degree in mechanical engineering. After brief stints with Baldwin, Cramps Shipyard, and the Dickson Locomotive Works, he joined the Schenectady Locomotive Works in 1900. After Schenectady had become part of the American Locomotive Company (Alco), his responsibilities included positions as Chief Draftsman, Assistant Mechanical Engineer, and Manager of the Electric Locomotive and Truck Department. In 1916, at age 43, Woodard left Alco for Ohio to become the Lima Locomotive Works' vice-president of engineering.

Woodard was convinced that the Mikado could be enhanced to preserve much of its starting strength while improving horsepower and fuel economy at higher speeds. New York Central president A. H. Smith echoed this belief, and he challenged the Lima engineers to come up with a 2-8-2 that would meet three clear objectives: 1) Deliver more power, for its weight, than any locomotive in the world; 2) Develop more power per ton of coal burned than any locomotive in the world; and 3) Be easier to operate than its predecessors, thereby improving safety and enabling reductions in terminal time. Woodard's efforts produced Michigan Central (NYC) number 8000, delivered on June 1, 1922, with 63" drivers, 28x30" cylinders, tractive power of 63,500 pounds (74,500 pounds with booster), and a total engine weight of 334,000 pounds. As the sole member of Lima order 1027 (shop number 6242), this hybridized 2-8-2 was tested by the Michigan Central in Detroit-Toledo and Toledo-Jackson (Michigan) freight service, and its front-end throttle, Elesco feedwater heater, Delta trailing truck, and expanded boiler capacity (compared to earlier Mikados) proved so successful a package that NYC promptly purchased 76 virtual copies along with the 8000 itself to create the H-10a class.

Even with the success of Lima 8000, Woodard and his colleagues and Lima continued to search for a more perfect balance of fuel economy, starting strength, and high-end horsepower. In June 1924, W. H. Winterrowd, Lima's Assistant to the President, disputed what his employer felt to be the contemporary perception that "the modern steam locomotive is inefficient, that it has reached a stage of development where further refinement cannot be offered, and that a point has been reached where it cannot meet practically and economically the traffic conditions of today. Nothing could be further from fact. The vitality of the steam locomotive is not on the wane. It is increasing. Development has not stopped. The possibilities of increased efficiency and capacity guarantee a continuing process. ...The greatest development of the future may be expected to take place in improvement of the locomotive as an efficient power plant."

Six months later, on January 24, 1925, Will Woodard observed, "Railway operating officers have for some time realized the importance of increased horsepower output from their locomotives as a factor in improving their operating conditions. We are coming to a time when railway men will begin to think about the horsepower of their locomotives instead of their tractive force, because, after all, it is the horsepower that does the work. We have reached the limit of the grate area which we can carry on a two-wheel trailing truck. We want more power and properly so. This means bigger grate areas. So we are ready for the next step, which is the four-wheel trailing truck and the big firebox."

By replacing the Mikado's two-wheel trailing truck with a four-wheel booster design with integral ashpan, Woodard's 2-8-4 was able to increase firebox dimensions and grate area (this wheel arrangement had been proposed by H. A. F. Campbell in October 1923, but in the form of a theoretical three-cylinder, 4,000-horsepower passenger locomotive that never left the drawing board). Woodard's design improved the locomotive's performance and power in the higher speed ranges by providing better drafting and more efficient heating of the large boiler, while the booster ensured that adequate reserves of starting strength were available. The large grate area decreased rates of combustion and reduced wear on the firebox, while the correspondingly larger ashpan design lowered maintenance costs by reducing the frequency of servicing. The use of limited cutoff improved fuel economy by using the expansive properties of steam to maximum advantage.

Woodard's theories were confirmed with the construction of Lima demonstrator A-1, delivered on February 11, 1925 (order 1070, shop number 6884). With stated objectives of developing "high horsepower capacity and improved economy in the use of fuel," the first 2-8-4 was quickly christened "Super-Power" by its builder, and the locomotive exhibited unmatched reserves of power during nationwide testing. It shared the 63" drivers and 28x30" cylinders of Lima 8000, but the engine weighed 51,000 pounds more than its predecessor and was measured to produce 76,800 pounds of tractive effort. The 2-8-4 had a massive 100 square foot grate area, compared with only 66.4 square feet on the 8000 (see Table 1). The A-1 underscored the validity of Woodard's emphasis on horsepower rather than tractive power, and on many occasions the A-1 was able to lift heavy trains up and over grades which had previously required doubleheaded Mikados.

This strength became particularly apparent during extensive tests on the Boston & Albany (NYC) main line between Selkirk (Albany), New York, and Washington,

Proudly displaying its builder's name, Woodard's A-1 launched the Super-Power era on February 11, 1925.
(Allen County Historical Society Collection)

(above, left to right) Mantis J. and Oris P. Van Sweringen.
(C&O Photo CSPR 10595YN; C&O Historical Society Collection)

(right) John J. Bernet, president of the Nickel Plate 1916-1926 and 1933-1935.
(C&O Photo CSPR 12333YN; C&O Historical Society Collection)

Table 1
Comparison of Lima A-1 and MC 8000

	A-1 (2-8-4)	8000 (2-8-2)
Cylinders	28 x 30	28 x 30
Cut-off in full gear	60%	Full stroke
Boiler pressure	240 lbs.	210 lbs.
Operating Weights:		
On drivers	248,200 lbs.	248,000 lbs.
On pilot truck	35,500 lbs.	29,000 lbs.
On trailer truck	101,300 lbs.	58,000 lbs.
Total engine	385,000 lbs.	335,000 lbs.
Driver diameter	63 inch	63 inch
Heating surfaces:		
Evaporative	5,110 sq. ft.	4,578 sq. ft.
Superheating	2,111 sq. ft.	1,780 sq. ft.
Grate area	100 sq. ft.	66.4 sq. ft.
Rated tractive force:		
Engine only	69,400 lbs.	66,700 lbs.
With booster	82,600 lbs.	77,700 lbs.
Factor of adhesion	3.58	3.72

Source: Railway Age, May 2, 1925.

Massachusetts. During a series of nine eastbound test runs accompanied by NYC dynamometer car X-8006 between March 28 and April 18, 1925, the new 2-8-4 produced from 23.5% to 33% more horsepower than comparison Mikado number 190, under identical operating conditions (the A-1's weight on drivers was just 200 pounds greater than the Mike's). The A-1 made such easy work of the winding grades over the Berkshire hills – and such an impression on B&A that the road ordered 25 copies within the year – that the editors of *Railway Age* suggested on August 8, 1925, that the nickname "Berkshire" be bestowed upon the new 2-8-4 wheel arrangement.

Woodard, Winterrowd, and Lima had led railroading into the Super-Power era, and the railroads were eager to adopt the new technology. Lima's contemporary trade advertising admonished the industry to break from the past by getting rid of "road-clogging" drag locomotives and embracing a new era in motive power. The builder extolled the numerous benefits of Super-Power, claiming the concept could yield a 33% increase in train load; a 17% increase in train speed; a 40% increase in gross ton-miles per train-hour; an 18.5% reduction in fuel consumption; and a 35.2% reduction in locomotive maintenance. Woodard produced research showing that his "modern" 2-8-4 design produced 76.5% more power at 25 mph than a drag-era 2-8-0. Lima knew these figures would be difficult for the railroads to ignore, and, sure enough, many came knocking.

Notable among these was the Erie, under the presidency of John J. Bernet. The promise of Super-Power was just what he needed to break the Erie of its drag-era shackles and turn it into a viable Chicago-New York competitor. Bernet had been appointed the Erie's president in 1926 by the Van Sweringen brothers, Cleveland developers who became railroaders by virtue of their 1916 purchase of the Nickel Plate from New York Central. NYC had been forced to sell the NKP in the wake of anti-trust action, and when the purchase closed the "Vans," as they were widely known, had also convinced former NYC vice-president J. J. Bernet to run their new railroad. An astute manager, Bernet made swift changes to the Nickel Plate's operations and physical plant that enabled the Road to speed up its freight trains and hold its own against its former parent, as well as expand into new territories through merger (LE&W and Clover Leaf) and stock control (C&O and Pere Marquette). Neglected under NYC domination, the Nickel Plate had been returned to solid footing – and the main architect of its renaissance was summarily moved east to work the same magic on the Vans' money-losing Erie.

Bernet's rejuvenation of NKP's locomotive roster with competent Mikados predated Super-Power's debut, but this was not the case on the Erie. He recognized in Super-Power a key solution to his new charge's problems, and in 1927-28 embraced the 2-8-4 wheel arrangement by purchasing a total of 105 engines from the Big Three (25 from Alco, 45 from Lima, and 35 from Baldwin). These were monstrous Berkshires, dwarfing the A-1 in many respects, with 70" drivers, 28-1/2x32" cylinders, engine weights of up to 461,470 pounds, and tractive power (with booster) of up to 72,000 pounds. During the first six months of 1928, Bernet saw his road move 13.4% more cars and 10% more gross ton miles per train; move 17.4% more gross ton miles per train hour; and operate trains 6.1% faster than in the same period one year before. By retiring 367 obsolete locomotives in 1927 and replacing them with the 2-8-4s and thirty new switchers, Bernet set into motion the means to reduce Erie's 1928 operating expenses 5% below 1927 levels and increase net operating income by 29%. The Erie's Berkshires were a watershed in 2-8-4 design, and although they proved slippery when starting a heavy train, their speed succeeded in returning their owner to solvency. Initially assigned between Meadville, Pennsylvania, and Marion, Ohio, the territory expanded as additional Berkshires, 125-car sidings, and strengthened New Jersey bridges entered service. In their prime, Erie's Berkshires operated from Marion to Secaucus, New Jersey, over all mainline districts with a grade of 0.2% or more

Even as Bernet was reviving the Erie, the Van Sweringen railroads were putting an organization in place which would facilitate improved equipment standards and benefit all participants. The Advisory Mechanical Committee (AMC) was created in 1929 to act as a central design bureau for the Erie, C&O, Pere Marquette, and Nickel Plate. In this role, under the direction of the C&O's William Black, the AMC coordinated the designs and specifications of everything from locomotives and passenger cars to stations, signage, and spitoons.

One of the AMC's first design projects was the Chesapeake & Ohio T-1 2-10-4 of 1930 – at the time, the largest two-cylinder steam locomotive in the world. Forty of these huge Super-Power engines (C&O 3000-3039) were built – appropriately enough, by Lima – to haul 15,000-ton coal trains single-handedly between Russell, Kentucky, and Toledo, Ohio. Black's designers and engineers (led by C&O motive power superintendent W. F. Barton) took Woodard's work to an even higher level, and avoided the shortcomings that had become apparent on some of the Super-Power locomotives designed and built during the previous five years. After borrowing Erie Berkshire 3377 (Baldwin, 1928) for dynamometer tests in May 1929, they settled on 69" drivers and 29x34" cylinders, and pro-

Lima got right to the point in promoting Super-Power.
(Allen County Historical Society; Author's Collection)

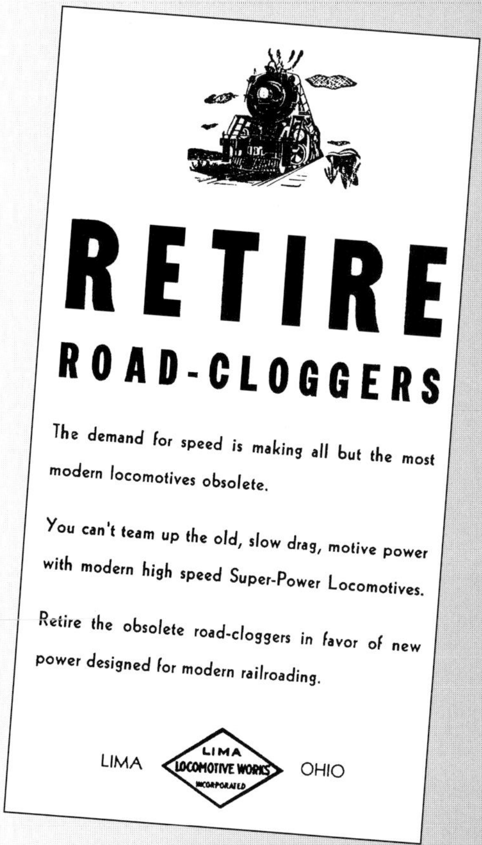

(top, and facing page) Chesapeake & Ohio T-1 of 1930 established the AMC "look" that would endure through almost two decades of steam production.

(Both, C&O Historical Society Collection)

(above) Nickel Plate S-2 reflected the AMC's strong influence.

(Lima Locomotive Works; Allen County Historical Society Collection)

(left) William Black, C&O Assistant to the President (Mechanical) and Advisory Mechanical Committee officer, led the AMC's T-1 project.

(C&O Photo; C&O Historical Society Collection)

(below) Had this January 31, 1934, Advisory Mechanical Committee proposal been adopted, the Nickel Plate's S class would have beaten New York Central's *Commodore Vanderbilt* 4-6-4 for the honors of being America's first streamlined steam design.

(Copyright 1962 Kalmbach Publishing Company, reprinted with permission from the October 1962 issue of TRAINS magazine)

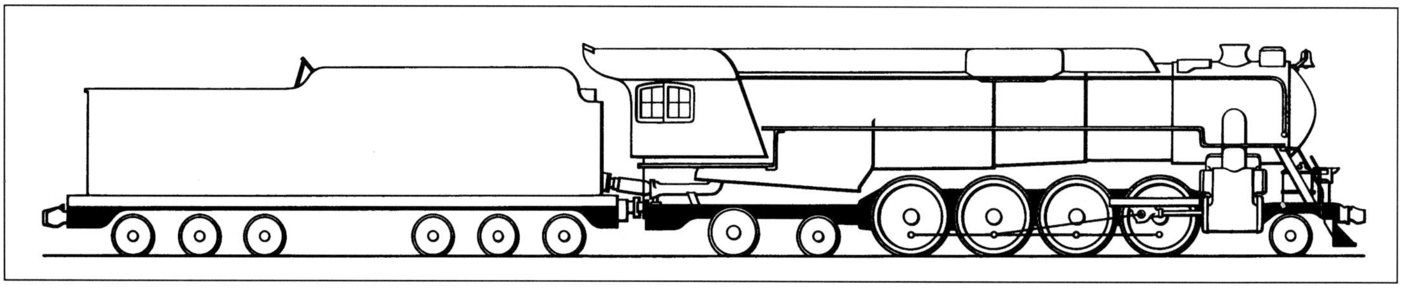

duced a locomotive with a staggering 106,584 pounds of tractive force with the booster cut in.

John Bernet's accomplishments in rejuvenating the Erie with Super-Power Berkshires came back to haunt him when he returned to the Nickel Plate presidency in 1933 (while still retaining the presidency of C&O and PM, which he had held since 1930). Even at their best, NKP's obsolescent Mikes couldn't maintain the Road's reputation for speed in the face of the new Super-Power competition – and the Erie was but one of the lines vying for business between Chicago and the East. The New York Central, Wabash, C&O, B&O, Pennsylvania, and Grand Trunk Western also wanted their piece of the pie.

In 1933, Bernet charged William Black and the AMC with the task of designing a Super-Power locomotive which would restore the Nickel Plate's competitive position. Building on the success of their T-1, the Committee's mechanical engineers designed a Berkshire that retained the 2-10-4's 69" drivers and 34" stroke. They maintained the T-1's 4.07 factor of adhesion by producing a new locomotive that was exactly 70% as heavy, with exactly 70% as much tractive power, as the T-1. The AMC acknowledged the Nickel Plate's shorter trains, flatter terrain, and resulting faster operation by deleting the T-1's booster. (Some NKP Mikados were booster-equipped, but the Road's motive power officials were never completely happy with the reliability and maintenance of these appliances.)

Super-Power, in perhaps its finest incarnation, had come to the Nickel Plate.

Beyond the "missing" pair of drivers, the external resemblance between the C&O and "scaled-down" NKP engines (as eventually built) was striking – but this was not to have been the case as late as January 31, 1934. Despite the AMC's mandate of standardization, the Committee proposed a skyline casing, sharply raked cab, and bullet nose for the Nickel Plate Berkshires that would have made them the first streamlined steam locomotives in the United States. The proposal was shelved, however, before the first NKP Berkshires went into production – an unnecessary expense for the historically thrifty Nickel Plate – and the strong "family" appearance of AMC-designed steam locomotives was thus established. The pilot-mounted headlight, an AMC design feature not adopted on the Nickel Plate's S class production, later became a Chesapeake & Ohio standard.

Nickel Plate motive power officials had also evaluated the merits of roller bearings between July 1 and July 11, 1931, when Timken 4-8-4 demonstrator number 1111 (the so-called *Four Aces*) conducted 12 Buffalo-Bellevue and Lima-Tipton, Indiana, test runs. The Northern, built for Timken by Alco in 1930 and equipped with roller bearings on all axles as a test and demonstration platform, ran 2,110 miles in NKP service before it was interchanged at Bloomington, Indiana, with the Chicago & Alton.

Erie Railroad S-2 Class 2-8-4 (series 3325-3349), built by Lima Locomotive Works in 1927.

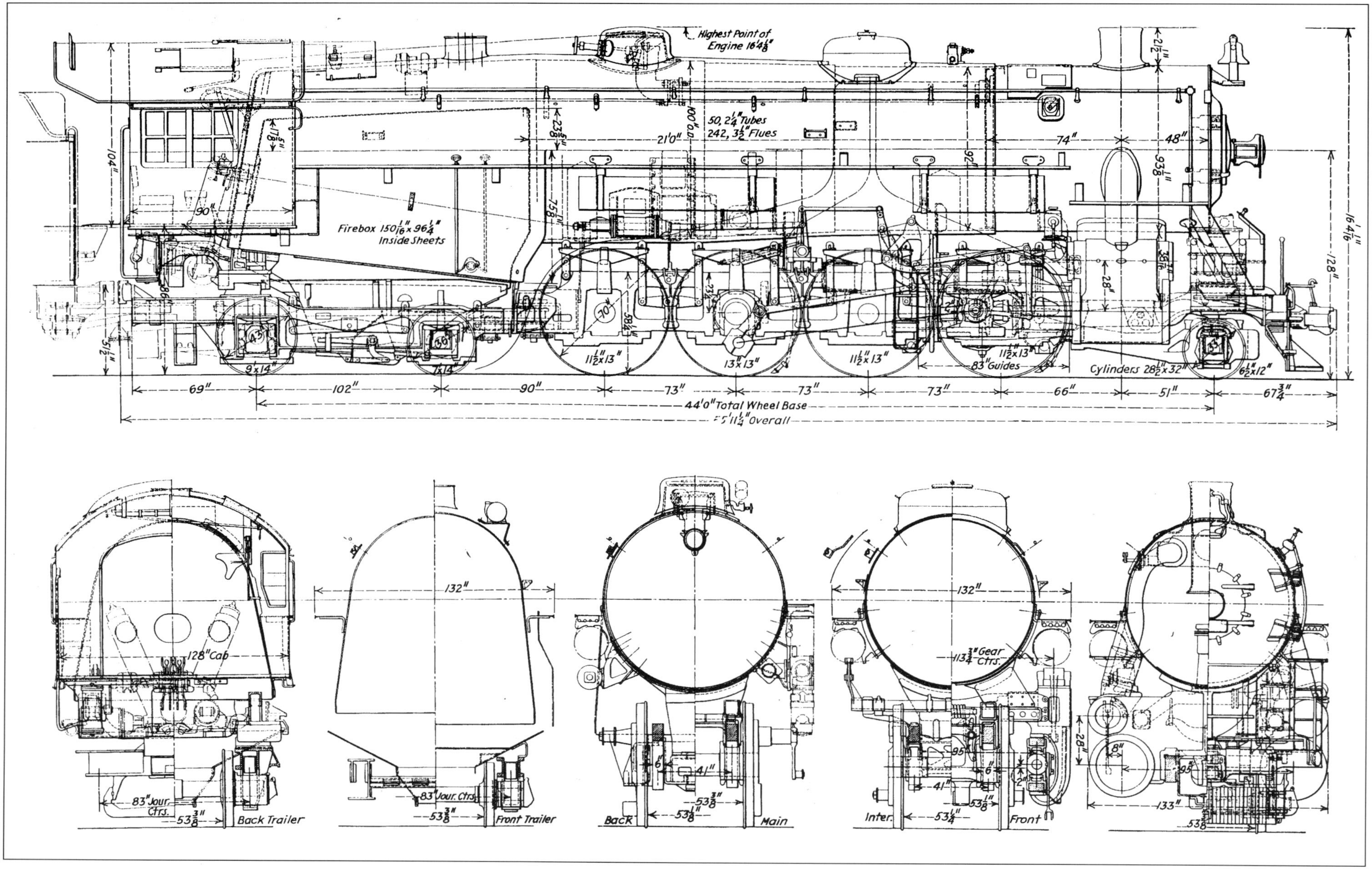

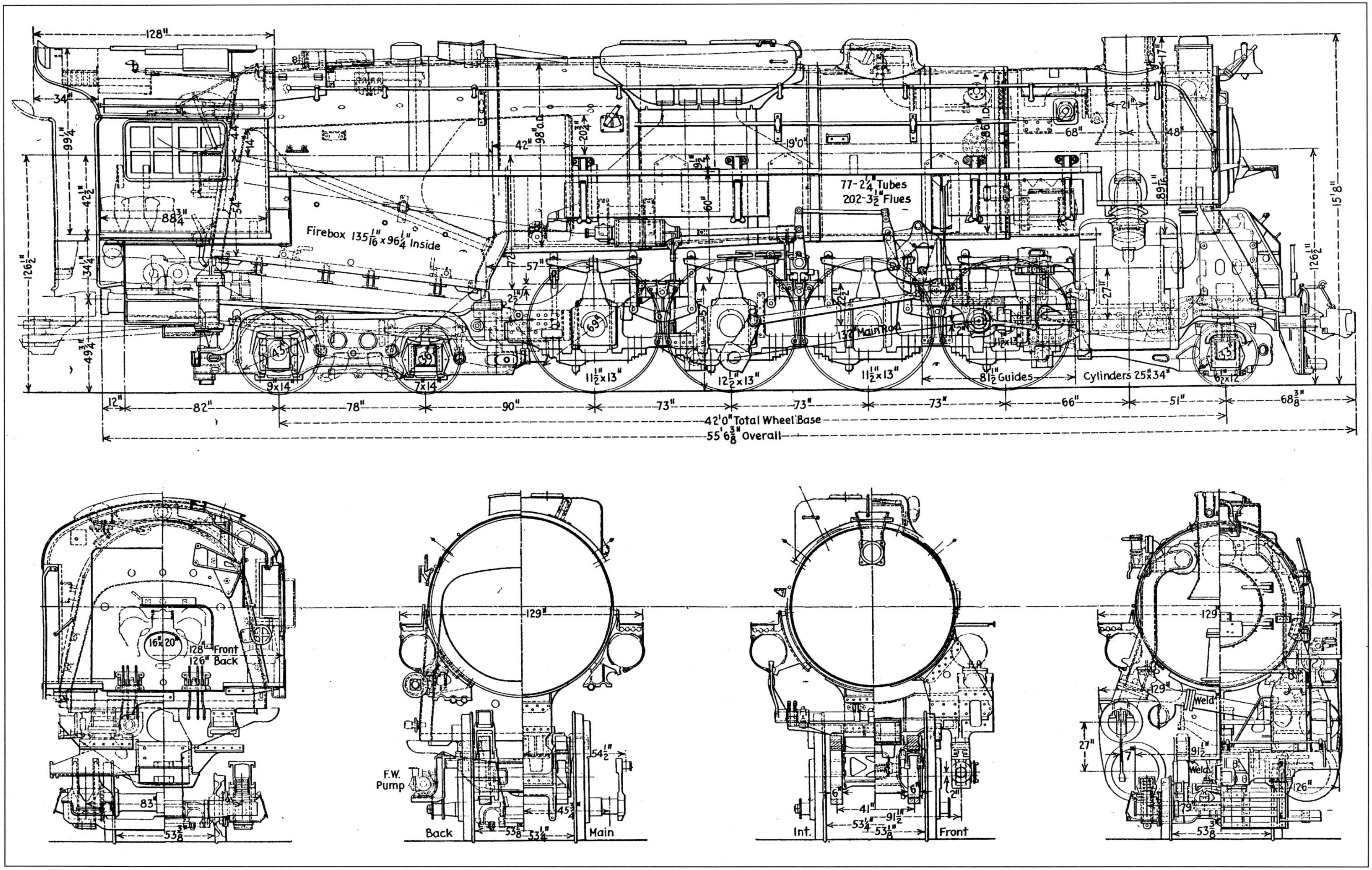

Nickel Plate Road S Class 2-8-4 (series 700-714), built by Alco in 1934.

THE S CLASS:

Chapter 2

In an era before telemetry, seven members of the test crew minister to the needs of brand-new Nickel Plate Road 700, reading gauges and checking cables strung to the trailing dynamometer car during a stop at St. Marys, Ohio, in September 1934. Sheet-metal pilot windbreak protected crew members taking cylinder readings during test runs.
(Clyde Helms; John B. Corns Collection)

One of the greatest ironies in the story of the Nickel Plate's Berkshires has to do with the builder of the initial 15 locomotives. Lima's origination of the Super-Power concept, its involvement in the AMC T-1 project, and its location on the Nickel Plate notwithstanding, the 1934 order for NKP 700-714 went to Alco in Schenectady. Ever frugal (and even moreso in the midst of the Depression), John Bernet had been obliged to ask Alco, Lima, and Baldwin to bid on the order, since the locomotives were among the Nickel Plate equipment Federally funded by a February 1934 Public Works Administration

DEBUT OF THE BERKSHIRES

loan. For its part, Lima had been forced by the depression to close its doors shortly after the T-1 project, and did not outshop its next locomotives until the spring of 1934. Alco's $1,345,856.25 bid clinched the order at an average cost of just under $90,000 per engine, while Lima was able to take some comfort in two existing Nickel Plate orders for five 0-8-0 switchers (order 1129, NKP numbers 300-304) and 25 extended-range 21-RA Mikado tenders.

By the time Alco began work on Nickel Plate order S-1736, there were already 282 Berkshires in service on American railroads, as well as two in Canada (see Table 2). There had been a five-year hiatus since the last Berkshires were built (by Lima, for Boston & Maine and Missouri Pacific), and Lima took advantage of the trade press' interest in the Alco/NKP order to remind the industry that both the Super-Power concept and the 2-8-4 itself had originated in Ohio, and not in Schenectady or Eddystone.

The entire Nickel Plate S class was delivered from Schenectady between September and November 1934. With specified engine operating weights of 416,000 pounds each, the Nickel Plate's new Berkshires were forced to

Table 2
Production Of 2-8-4 Locomotives, 1925-1934.

Owner	Builder	Qty	Date	Engine Weight	Cylinders	Drivers	Pressure	T.F. (lbs)
Lima	Lima	1	1925	385,000	28x30"	63"	240 psi	82,600 *
B&A	Lima	55	1925	389,000	28x30"	63"	240 psi	69,400 *
IC	Lima	50	1926	388,000	28x30"	63"	240 psi	69,400 *
C&NW	Alco	12	1927	397,000	28x30"	63"	240 psi	67,200 *
AT&SF	BLW	4	1927	396,500	27x32"	63"	220 psi	69,200 *
Erie	Alco	25	1927	443,000	28-1/2x32"	70"	225 psi	70,000 *
Erie	Lima	45	1927	457,500	28-1/2x32"	70"	250 psi	72,000 *
Erie	BLW	35	1928	461,470	28-1/2x32"	70"	225 psi	71,000 *
B&M	Lima	25	1928-9	393,000	28x30"	63"	240 psi	69,400 *
TH&B	MLW	2	1928	383,000	28x30"	63"	240 psi	69,000
MP	Alco	5	1928	404,000	28x30"	63"	240 psi	70,500 *
MP	Lima	25	1929	412,200	28x30"	63"	240 psi	69,400
NKP	Alco	15	1934	416,000	25x34"	69"	245 psi	64,100

* With booster

Sources: Railway Age, November 17, 1934; Lima Locomotive Works Delivery Records.

take a circuitous route to reach home rails, traveling via D&H to Binghamton, New York, and then over DL&W to Buffalo and NYC to Erie, Pennsylvania, in order to bypass a light Nickel Plate viaduct near Dunkirk, New York (in fact, Berkshires would not reach Buffalo on their own rails until several of the Nickel Plate's East End bridges were strengthened during World War II). Total weight on drivers was 254,000 pounds, and the loaded tender weighed 358,000 pounds.

The first five engines, NKP 700-704, were assigned to operate between Cleveland, Lima, and Frankfort. Class engine 700 was the first Berkshire into Frankfort, Indiana, when it arrived with train 49 over the old LE&W from Lima on September 13, 1934. It returned to Lima the next day on train 90, ushering in a new era on the Nickel Plate.

The other ten Berkshires, NKP 705-714, were equipped with Automatic Train Stop (ATS) equipment to permit operation between Chicago and Bellevue (all locomotives operating over the Nickel Plate's Chicago Division were required to be ATS-equipped). Reflecting their extra hardware, the ATS-equipped Berkshires had a unit cost approximately $550 higher than their non-ATS counterparts.

In advance of the Berkshires' deployment on the Sandusky Division, speed restrictions across six bridges were revised in July 1934 to accommodate these largest-ever Nickel Plate locomotives (as well as the H-6 class Mikes retrofitted with new larger tenders just delivered from Lima). The weight of the new locomotives was of sufficient concern that superintendent C. R. Wright issued this letter to his roadmasters on September 7, 1934:

"The first Class 'S' engines have been received on our line and will soon be put in service between Bellevue and Frankfort. These engines are numbered 700 to 714 inclusive.

"These engines are heavier than any engine that has operated on the railroad. When these engines are put in operation all of the foremen and men between Arcadia and Frankfort should make careful inspection of track, roadway, and structures and report promptly any evidence of weakness which may develop in track or bridges.

"If any of the men are not familiar with the bulletin covering the speed restrictions governing the operation of these engines they should be given this information and instructed to report any violation of the speed restrictions by these engines.

From boiler to backhead, Nickel Plate Road 700 takes shape at the American Locomotive Company's Schenectady, New York, works in the summer of 1934.

(top) The firebox and boiler have been moved outdoors for this photograph, en route to the erecting hall.

(above) As construction progresses, the Worthington Type S feedwater heater, air reservoir saddles, multiple bearing crosshead and guide, and cylinders are in place, and some of the short-lived firebox sheathing has been applied.

(right) Cab interior and backhead of NKP 700, just prior to delivery in September 1934. Of note are reverse wheel at right and five Simplex stoker steam jet pipes at left.
(Alco Photos; John B. Corns Collection)

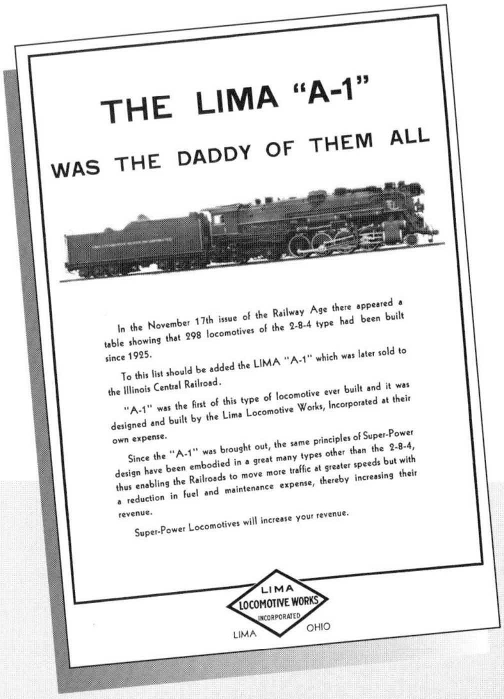

(left) The S class order may have gone to Alco, but Lima got the last word in November 1934 with trade advertising reminding the industry just where Super-Power had originated.
(Allen County Historical Society; Author's Collection)

(below) Hard feelings aside, the Alco Berkshires were handsome machines reflecting the strong styling influence of the Advisory Mechanical Committee and their C&O T-1 2-10-4 of 1930. Patriarch 700 hasn't yet received its pilot windbreak in this September 1934 view, but NKP dynamometer car X-50041 is coupled behind the tender nonetheless.
(John B. Corns Collection)

(bottom) Polished cylinder head covers and firebox sheathing were as-delivered hallmarks of the Alco Berkshires, as on brand-new 705 at St. Marys, Ohio, on October 9, 1934.
(Clyde E. Helms; John B. Corns Collection)

702
702

"I do not know that these engines will be operated on trains doing local switching work but if they are used for this service it may be necessary for you to place additional restrictions against the use of this power on certain tracks."

The S class set the pattern for subsequent orders, so it's worthwhile to take a closer look at the class' specifications. The 44' 8-7/16"-long boiler was designed to operate at a pressure of 245 pounds, at which point 64,100 pounds of tractive power were produced. As a contingency, this first NKP order had provision to increase boiler pressure to 250 pounds, and to increase the diameter of the cylinders from 25" to 26". Provision was also made for application of a booster, but this was also never done. The ratio of boiler horsepower to cylinder horsepower, according to NKP mechanical records, was 114.9 (fractionally less than the C&O T-1). The new Berkshires' factor of adhesion was identical, though, at a healthy 4.07.

The 135-1/16" x 96-1/4" firebox afforded a 90.3 square foot grate area, fed by an automatic stoker. A pair of thermic syphons improved boiler circulation and added 98 square feet of heating surface as they enabled colder water to be drawn upward by convection through two firebox tubes. The Elesco Type E superheater had 1,992 square feet of heating surface. Total evaporative heating surface was 4,818 square feet. The engines were equipped with Worthington Type S feedwater heaters, Nathan Simplex NL injectors, and a Nathan low water alarm. The 116"-long smokebox was capped with a 21"-diameter stack, 15'-8" above the railhead. The stack casting incorporated a separate exhaust flue for the feedwater heater. Like the C&O T-1, a visored headlight, with number indicators, was centered on the smokebox door, and a cord-operated bronze bell hung in a cradle ahead of the feedwater heater. Two shielded New York Air Brake Company 8-1/2" cross-compound air pumps were placed above the bar-stock pilot.

Frames were assembled from cast and fabricated steel components. The General Steel Castings engine truck incorporated friction bearings and 33" rolled steel wheels. Drivers were 69" diameter, and the Delta trailing truck

Flying green sheet metal "flags," Nickel Plate 712 leads First 68 (a Peoria-Buffalo time freight) across the Pennsylvania Railroad diamond and east out of Frankfort yard at 10:00 am on August 13, 1945. As with the other members of this class, 712 has lost its cylinder head covers and firebox sheathing. *(Richard J. Cook, Sr.; Allen County Historical Society Collection)*

712
712

had 36" rolled steel (front) and 43" steel-tired (rear) wheels. Baker long-travel valve gear was controlled by Franklin precision reverse gear.

Cabs featured wood sash and insulation, and the water-bottom 21-RE tenders rode on six-wheel Buckeye trucks with 33" wheels and friction bearings. Tender capacity was 22 tons of coal and 22,000 gallons of water. Roller bearings had been proposed for the S class but, like the streamlined shrouding, were rejected in the interests of economy.

The Road's corporate initials (NYC&STL) appeared in 2-1/2" imitation gold letters on both sides of the Berkshires' large rectangular sandboxes, which delivered sand to the drivers through four pipes per side. Cab-side 9" road number and 1" class/tractive power lettering was also imitation gold, while running board edges and driver tire rims were accented in white. Tender lettering was imitation gold, with 2-1/2" corporate initials on the collars, 9" script roadname on the sides, and 2" coal/water capacities and 9" roadnumbers on the rear.

Even 15 new Super-Power Berkshires were not enough to fill the motive power vacuum created by John Bernet's scrapping of 156 obsolete locomotives during the early 1930s. Many of these were ex-Clover Leaf and LE&W antiques surplus to Bernet's more efficient operation, but additional engines were still required to move the Nickel Plate's manifests and maintain the higher levels of service Bernet had introduced. Since funds for additional purchases were not available, a total of fifteen Mikados were leased between 1941 and 1943 from Lackawanna, Pere Marquette, and C&O, along with two C&O G-7 class 2-8-0s used at Bellevue. This proved to be an acceptable solution to the Nickel Plate's motive power shortage, at least in the short term, but all of the rules changed on December 7, 1941.

Like the rest of the country, the Nickel Plate was at war.

(left) Sporting the AMC-influenced "flying" numberboards applied to Nickel Plate steam after 1944 (and also used by C&O), 709 cuts through Elwood, Indiana, at 40 mph with an 83-car westbound freight on May 29, 1949.

(Richard J. Cook, Sr.; Allen County Historical Society Collection)

(above) Under a clean stack, 707 negotiates the east end of a seven-mile mainline track relocation at Oakland, Ohio, in 1948. Part of John W. Davin's postwar infrastructure improvements, the 1947 project reduced curvature and grades west of Bellevue.

(NKP Photo; John B. Corns Collection)

(above right) The Nickel Plate's second-oldest 2-8-4 rests at Madison (East St. Louis), Illinois, on May 13, 1952. Bell and new numberboards combined to obscure the Worthington type S feedwater heater in this "intermediate" front-end appearance.

(Louis A. Marre Collection)

(right) Fireman's side of 702 in the intermediate configuration, prior to application of Mars light.

(John B. Corns Collection)

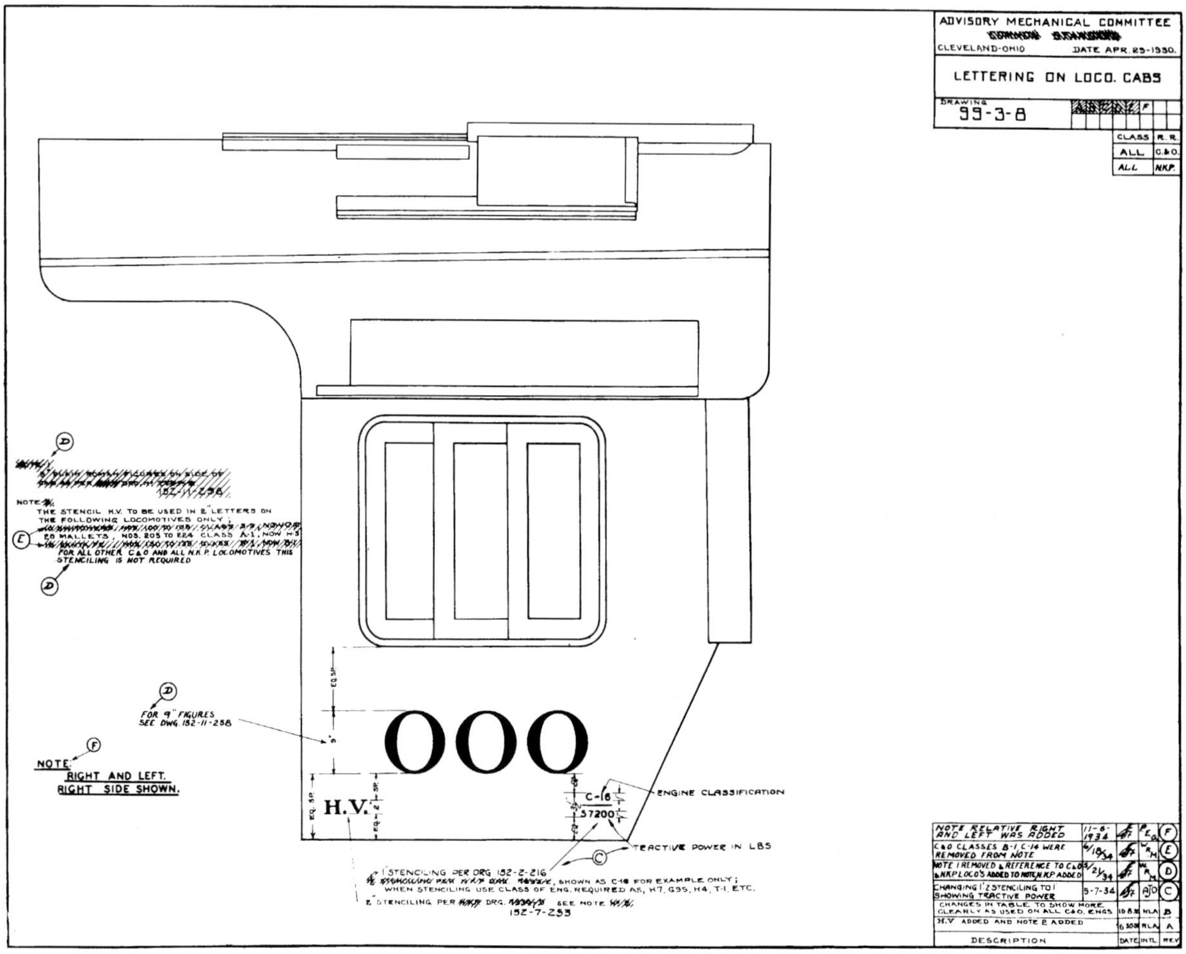

(Nickel Plate Road Historical & Technical Society Collection)

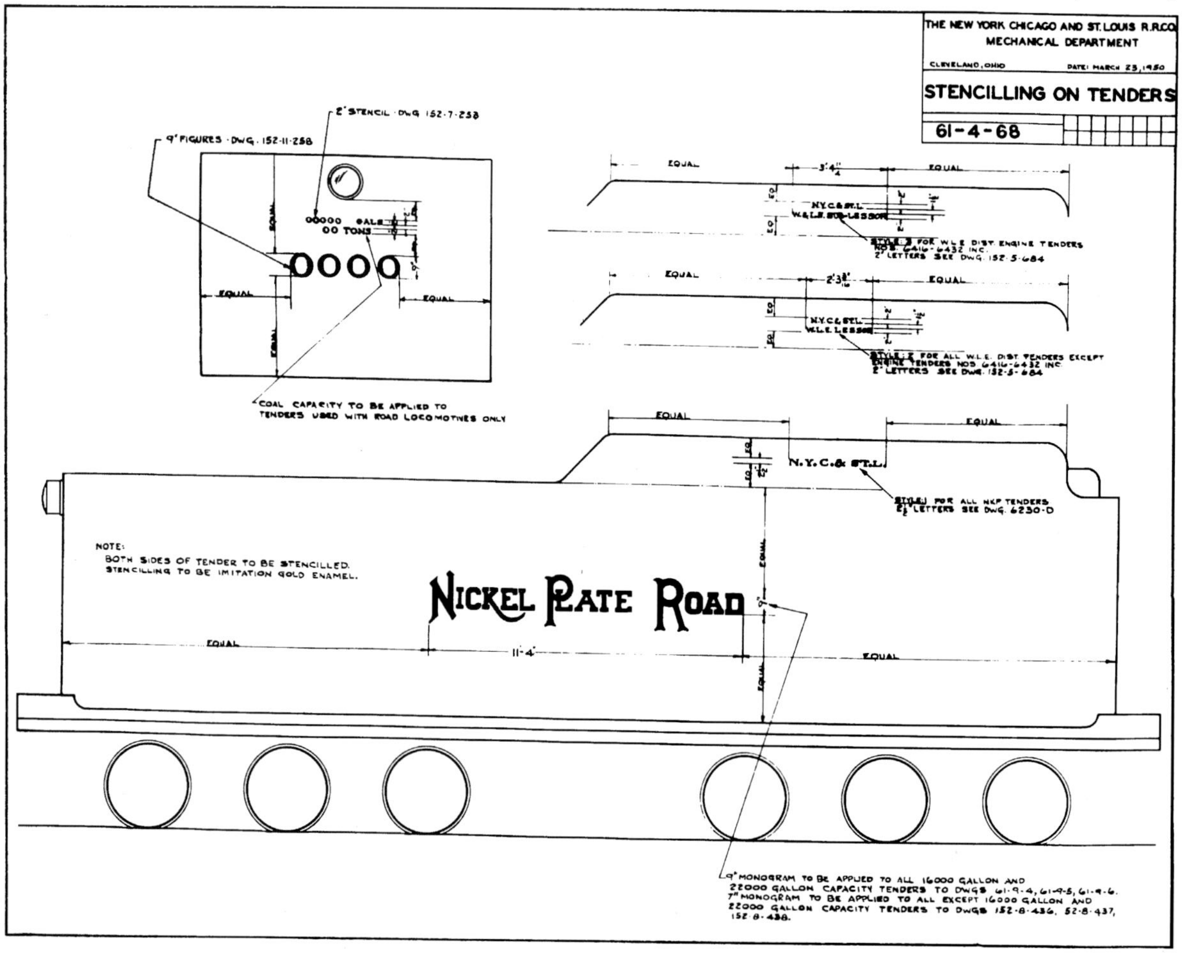

Stiff winds off Lake Erie ensure that 707's considerable smoke plume keeps pace as the Alco leads an eastbound extra through Angola, New York. The New York Central main line is visible at right – this was one of many locations between Buffalo and Cleveland where the two roads were within sight of each other.

(Jim Shaughnessy)

(above) Dusty and soot-covered in its twenty-first year of service, the 705 rests at Frankfort on February 24, 1955. Compare this view to the one on page 20 ... the engine has received the usual latter-day Mars light, turbogenerator, and numberboard modifications, as well as a retrofit Pyle-National aluminum headlight (identifiable by the replacement gabled number plate).

(Jay Williams, Jr., Collection)

(facing page) By the time the 706 led eastbound train 86 over the W&LE between Scio and Jewett, Ohio, in May 1957, it had become the last Alco S-class Berkshire in service. Note the unusual baffles behind this engine's footboards, a safety feature borrowed from the ex-Wheeling S-4 class Berks.

(John A. Rehor; John B. Corns Collection)

Shown here leading a work train over former W&LE track through Bellevue late in its career, the 710 was one of a dozen Nickel Plate Berkshires to be fitted with replacement plow-type pilots at various times after World War II. These were fairly crude, home-built steel plate affairs, with a shallow prow beneath the coupler. The others were 700, 703, 715, 722, 724, 727, 736, 741, 742, 744, and 745. Eighteen other NKP steam locomotives, mostly H-6 class Mikados, are known to have received plows beginning in the late 1930s, and some of these were applied to the above-noted Berkshires as the older engines were retired or transferred.

(John A. Rehor; John B. Corns Collection)

Roster of NKP S Class Berkshires

American Locomotive Company (Schenectady), Order S-1736, Sept.-Nov. 1934

Number	Builder's Number	Delivered	Disposition
700	68646	September 1934	Retired 10/57. Sold for scrap 4/61
701	68647	September 1934	Retired 8/56. Sold for scrap 6/57
702	68648	September 1934	Retired 5/57. Sold for scrap 6/61
703	68649	September 1934	Retired 9/56. Sold for scrap 5/57
704	68650	September 1934	Retired 8/56. Sold for scrap 5/57
705	68651	September 1934	Retired 6/56. Sold for scrap 8/57
706	68652	September 1934	Retired 11/57. Sold for scrap 8/63
707	68653	September 1934	Retired 9/57. Sold for scrap 4/61
708	68654	October 1934	Retired 11/57. Sold for scrap 4/61
709	68655	October 1934	Retired 8/57. Sold for scrap 4/61
710	68656	October 1934	Retired 8/56. Sold for scrap 6/57
711	68657	October 1934	Retired 4/57. Sold for scrap 11/61
712	68658	October 1934	Retired 10/57. Sold for scrap 4/61
713	68659	November 1934	Retired 5/57. Sold for scrap 11/61
714	68660	November 1934	Retired 7/56. Sold for scrap 6/57

Chapter 3

THE S-1 CLASS:

On June 1, 1942, NKP 718 takes shape on the floor of Lima's erecting hall as part of the "Loco's" construction order 1160. The S-1 Berkshires shared the use of fabricated frames with their Alco-built predecessors.
(Lima Locomotive Works; John B. Corns Collection)

Nickel Plate 715-729 were the first NKP Berkshires built by Lima, which had lost out to Alco in the bidding for the engines delivered as NKP 700-714 in 1934. That is not to suggest that Lima had lost any momentum in the construction of Van Sweringen/AMC Super-Power, however. In the wake of the 45 pre-AMC Berkshires delivered to Erie in 1927, Lima built the forty C&O T-1s of 1930, five C&O J-3 class 4-8-4 Greenbriers in 1935, and 15 N class Berkshires for C&O affiliate Pere Marquette in late 1937 (see Table 3).

The Nickel Plate's first 15 Lima Berkshires were delivered under the administration of George Brooke in June and July 1942, and were assigned NKP class S-1. Even as these locomotives were under construction, NKP

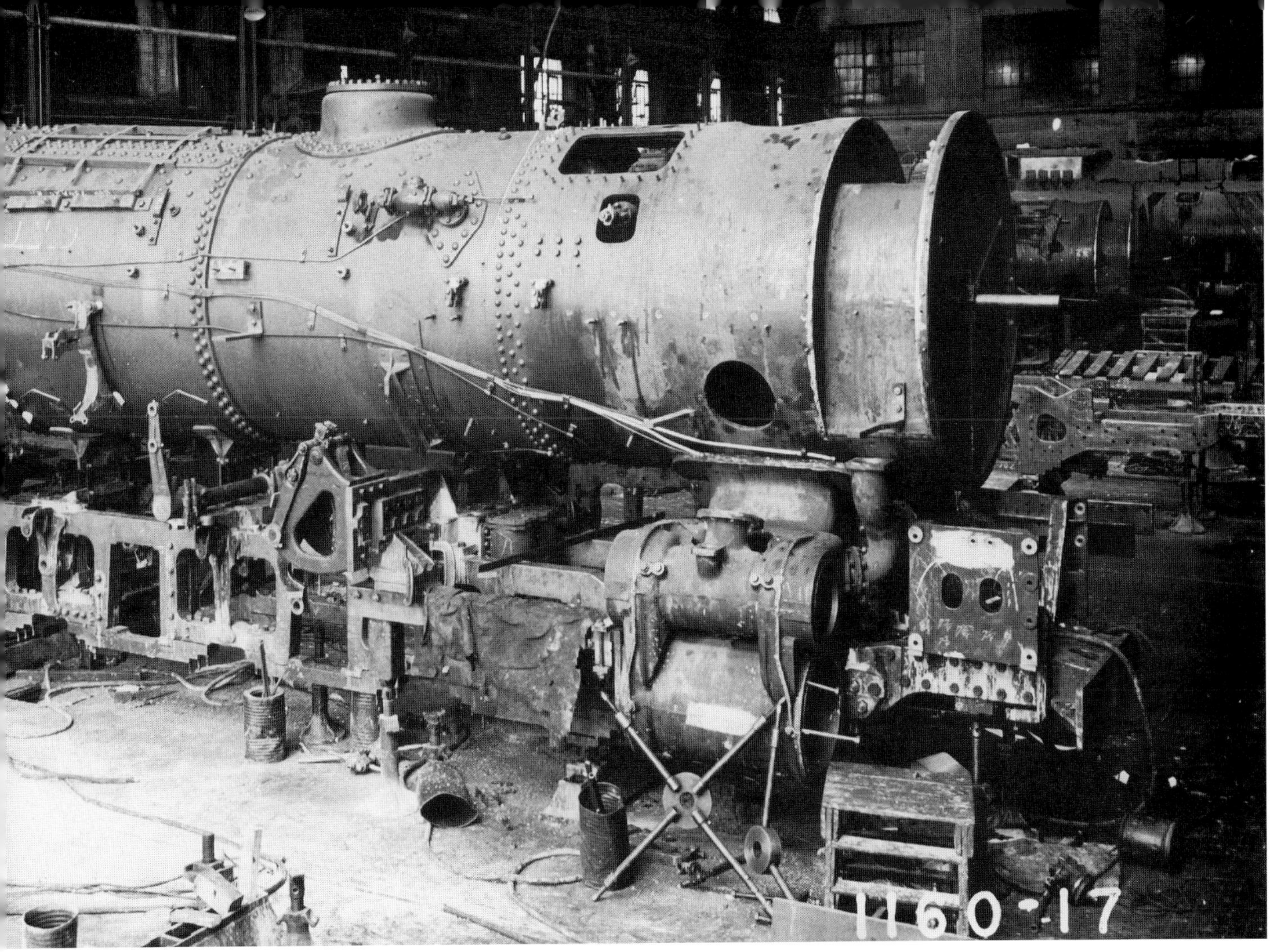

HOMECOMING

President Brooke appealed to his board on February 17, 1942, to authorize an additional $1.6 million for ten more S-1 Berkshires. Brooke's rationale, as recorded in the minutes of that board meeting, was urgently simple: "The President stated that the current volume of business was the largest in the Company's history. He reviewed the Company's motive power and equipment situation and called attention to the need of providing additional equipment to meet the increasing demands incident to the expansion of business attending the National Defense Program. He pointed out that it had been necessary to rent 17 engines ... which, however, were subject to recall by the owners on short notice, and reported that in view of the extreme need for additional motive power he had, after receiving the approval of three members of the Executive Committee, arranged for the acquisition by or on behalf of the Company of 10 2-8-4 freight locomotives, with 22,000 gallon tenders ... from Lima Locomotive Works, from whom 15 similar locomotives with tenders had previously been ordered." With the process already underway, the board's unanimous approval of Brooke's motion was little more than a formality, but the Road was obliged to wait until November 21, 1942, before the War Production Board authorized Lima to commence construction of NKP 730-739. John Dickinson, Lima's Assistant to the President, advised Brooke on that date that five of the locomotives were in the builder's February 1943 schedule and that Lima was "in hopes of shipping five in February and the other five in March." Even a year after the order had been approved by

continued on page 42

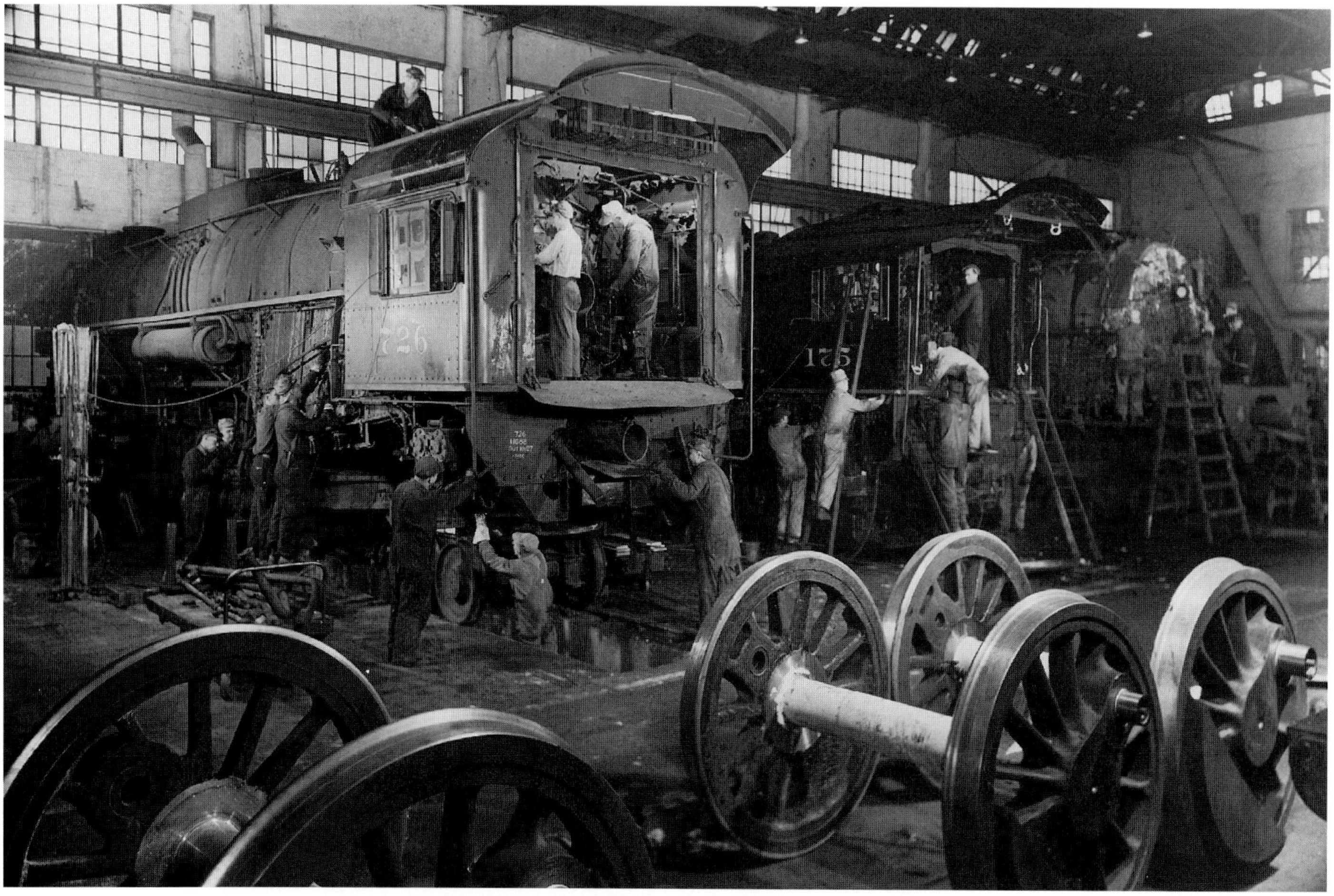

(above) Superior workmanship and strict adherence to maintenance schedules ensured that the Nickel Plate's locomotives could meet the rigors of World War II traffic levels as well as the postwar boom. In this 1945 view inside Conneaut's backshop, S-1 number 726 and another 2-8-4 undergo Class 3 repairs while the cab of Hudson 175 is readied for removal. The 726 rests on diminutive shop trucks, having been unwheeled to permit running gear repairs.

(NKP Photo by George J. Morris; John B. Corns Collection)

(left) This tender's-eye view of an S-1 Berkshire under construction at Lima reveals, from top to bottom, the crew grip storage rack; the relatively uncluttered backhead and controls; Butterfly firebox doors and engineer's heat shield; stoker connection; drawbar; and tender water connections.

(Lima Locomotive Works; John B. Corns Collection)

(facing page, top) Sporting new "flying" numberboards, 728 rests at South Lima, Ohio, on June 11, 1950, in the shadow of its builder. When these new illuminated numberboards were applied, original small boards flanking the headlight were blacked out. Installed, the new boards cost $121.00 per pair when the retrofit program began in July 1947.

(John B. Corns Collection)

(facing page, bottom) Form follows function. Canvas awning and gangway curtains, leather arm rest, glass wind deflector and floor-level air vent were about the only concessions to crew comfort, as exemplified on the Nickel Plate's youngest S-1 near the end of its career.

(Jim Shaughnessy)

728
NYC&StL
NICKEL PLATE ROAD
728
728

739

(above) S-1 number 718 passes beneath Ohio state route 299 (later route 13) as it leads a westbound train between Shinrock and Avery, Ohio, in June 1957. Note unusual painted number plate numerals.

(Philip T. Horning; John B. Corns Collection)

(facing page) The Buffalo and Cleveland Divisions comprised the Nickel Plate's "East End." The Road stayed within a few miles of the Lake Erie shore and was obliged to span a number of wide creek valleys on spectacular steel viaducts. The engineers who designed these late-nineteenth century structures never contemplated Cooper's ratings heavy enough to support the likes of a Berkshire, and it was not until World War II that traffic levels could justify the rebuilding and strengthening of these spans. Berkshires, like S-1 number 719 crossing Conneaut Creek eastbound in this mid-1950s view, entered service east of Cleveland in 1943.

(TLC Collection)

Table 3

AMC Super-Power Production by Lima Locomotive Works, 1929-1949 (Non-articulated road locomotives)

Road	Class	Series	Type	Order	Delivered
C&O	T-1	3000-3039	2-10-4	1122	8/29/30 - 11/24/30
C&O	J-3	600-604	4-8-4	1131	12/11/35 - 12/23/35
PM	N	1201-1215	2-8-4	1143	9/11/37 - 12/02/37
PM	N-1	1216-1227	2-8-4	1155	10/09/41 - 11/07/41
C&O	J-3-A	605-606	4-8-4	1156	1/29/42 - 1/31/42
NKP	S-1	715-729	2-8-4	1160	6/04/42 - 7/30/42
NKP	S-1	730-739	2-8-4	1166	3/18/43 - 5/29/43
NKP	S-2	740-754	2-8-4	1179	1/11/44 - 3/08/44
PM	N-2	1228-1239	2-8-4	1180	3/24/44 - 5/15/44
NKP	S-2	755-769	2-8-4	1184	8/14/44 - 9/28/44
C&O	K-4	2740-2749	2-8-4	1192	9/22/45 - 12/07/45
VGN	BA	505-509	2-8-4	1194	6/27/46 - 7/25/46
C&O	K-4	2750-2759	2-8-4	1193	1/10/47 - 3/05/47
C&O	J-3-A	610-614	4-8-4	1201	6/10/48 - 6/29/48
NKP	S-3	770-779	2-8-4	1209	3/14/49 - 5/13/49

Source: Lima Locomotive Works Delivery Records.

Doubleheaded Nickel Plate Berkshires were not commonplace, and when the 2-8-4s operated in tandem it was usually for one of two reasons.

(above) The NKP invariably handled more loads eastbound than westbound, and hence operated more eastbound trains to move the disproportionate tonnage. As a result, locomotives tended to "collect" on the East End, and westbound weekend doubleheaders corrected the imbalance. On very rare occasions, like this mid-1950s view at Cleveland's East 37th Street, eastbound doubleheaders also ran. With windows open, PCC car passengers take in the performance as S-1 number 718 and radio-equipped S-2 number 744 on train OB-2 confront ex-W&LE Berkshire 805 and an Alco S2 diesel switcher in the distance.

(right) While the strength of doubleheaded Berkshires was rarely needed on the main line, the opposite was often true on the Wheeling District. Heavy mineral traffic and the curving, hilly profile of the former Wheeling & Lake Erie combined to tax a single Berkshire's 64,100 pounds of rated tractive power. S-1 Berks 716 and 728 have only a wooden caboose in tow at Adena, Ohio, in 1957. Demoted from the Nickel Plate's other Districts by encroaching dieselization, these locomotives and their 800-series former W&LE roster mates had just a few months left before Alco RSD12s and EMD SD9s took over in the Ohio coal fields.

(2 NKP Photos by Willis A. McCaleb; John B. Corns Collection)

716
716

(above) The South Lorain turn, with S-1 number 731 in charge, is about to cross the Baltimore & Ohio's Sandusky-Newark, Ohio, line as it passes the junction station at Kimball on Labor Day weekend 1957. "HI" was the telegraph call sign for Kimball, which was 7.3 miles east of Bellevue on the Cleveland Division main line.
(Don Wood)

(below) The hilly terrain and omnipresent hoppers are clues that the 729 was another of the S-1 Berkshires to be bumped to the Wheeling District as diesels took over elsewhere on the Nickel Plate. With ex-W&LE steel caboose 709 in tow, the grimy 2-8-4 negotiates yard trackage at Adena in the spring of 1957.
(NKP Photo by Willis A. McCaleb; John B. Corns Collection)

(right) The 737 heads a string of Wheeling District coal hoppers through Harmon, Ohio, in the mid-1950s.
(NKP Photo by Willis A. McCaleb; John B. Corns Collection)

(below) Another view of 735, this time passing eastbound through Avon, Ohio, on February 23, 1958. Three weeks later, on March 16, the engine was retired at Conneaut.
(Philip T. Horning; John B. Corns Collection)

735
O COR
735

(left) In a view from a coach vestibule of NKP Buffalo-Chicago train 7, the *Westerner*, S-1 number 735 leads eastbound freight CB-2 through the Fostoria, Ohio, station in 1957.
(NKP Photo by Willis A. McCaleb; John B. Corns Collection)

On July 19, 1956, Extra 723 East held the main at Perry, Ohio, as an opposing freight behind S-2 number 744 took the siding.
(Don Wood)

Back in its hometown, the 732 goes for a spin on South Lima's 90-foot turntable – just long enough to handle the S-1's 87′ 8-3/4″ wheelbase. Take particular note of the 722's tender in this post-1951 view. Still equipped with wartime steam and signal lines for passenger service, it also has had a second backup light installed – a dual sealed beam unit up on the tank deck.
(John A. Rehor; John B. Corns Collection)

the Nickel Plate board, Dickinson's forecast proved somewhat optimistic. Construction order 1166 was finally delivered between March 18 and May 29, 1943.

Even with the unprecedented demands of wartime traffic, and despite the fact that NKP had been obliged to wait 13 months for delivery of Berkshires 730-739 to begin, these ten locomotives were not placed in service immediately. Instead, they were sent to the NKP's Lima shops as soon as they were delivered. Amazingly, they were then unwheeled for a thorough lubrication and inspection of brakes and running gear, incurring between 50 and 79 man-hours of labor per locomotive before being released for revenue service.

(above) S-1 number 733 is about halfway between Buffalo and Conneaut as it leads a westbound through the three-chute, 300-ton coaling station at Pomfret, New York, in 1957. Nemesis New York Central's main line is visible at left.

(Jim Shaughnessy)

(facing page, top) St. Louis (Madison)-Buffalo manifest KC-44 leaves Tipton, Indiana, behind the 724 in this post-1950 scene.

(NKP Photo by Willis A. McCaleb; John B. Corns Collection)

(right) Extra 723 East approaches Conneaut on July 19, 1956, en route to Buffalo and eastern connections.

(Don Wood)

724

723
723
723

All 25 S-1 Berks were essentially copies of the Alco-built S class – hardly surprising where the AMC was involved – right down to their fabricated frames. All were equipped with ATS gear. Unlike the Road's later Berkshires, which incorporated solid cast steel beds with integral cylinders, Nickel Plate 700-739 gained reputations as rough riders. Subtle differences from the S class included a modified sander arrangement, and adoption of a pneumatic bell ringer instead of a manual cord. The two S-1 orders were marginally heavier than their predecessors, by just 600 pounds for the engine alone. Total S-1 operating weight with tender was 791,200 pounds, or 4,300 pounds more than an S class engine. The only other significant differences were a 4.02 factor of adhesion, an improved Worthington 5-SSA feedwater heater, and heavier class 22-RG tenders.

The arrival of the first 15 of these Nickel Plate "War Babies" sent the 17 leased locomotives home, but not soon enough to save C&O Mikado 2314, which had been destroyed by a boiler explosion at Red Key, Indiana, on March 31, 1942, with the loss of three crewmembers' lives. Engineer E. T. Benner, Fireman W. H. Hershberger, and Brakeman J. E. Schletter perished when the Mike's boiler was blown over 330 feet clear of the engine. C&O 2345 had also been wrecked while on lease to the Nickel Plate – at Coldwater, Ohio, just days before the explosion of 2314. Delivery of the S-1 class Nickel Plate Berks also led to revised assignments for the original S class engines. By mid-1943, NKP 700-719 were spending most of their time on the ex-LE&W lines, while 720-739 were assigned to the Chicago and Fort Wayne Divisions.

Roster of NKP S-1 Class Berkshires

Lima Locomotive Works, Construction Order 1160, June-July 1942.

Number	Builder's Number	Delivered	Disposition
715	7860	June 4, 1942	Retired 10/57. Sold for scrap 4/61
716	7861	June 6, 1942	Retired 9/57. Sold for scrap 2/58
717	7862	June 12, 1942	Retired 9/57. Sold for scrap 2/58
718	7863	June 15, 1942	Retired 8/60. Sold for scrap 12/63
719	7864	June 17, 1942	Retired 8/60. Sold for scrap 1/64
720	7865	June 22, 1942	Retired 8/60. Sold for scrap 10/63
721	7866	June 25, 1942	Retired 8/56. Sold for scrap 6/57
722	7867	June 29, 1942	Retired 6/56. Sold for scrap 8/57
723	7868	July 7, 1942	Retired 7/56. Sold for scrap 8/57
724	7869	July 16, 1942	Retired 12/57. Sold for scrap 1/62
725	7870	July 23, 1942	Retired 8/60. Sold for scrap 8/63
726	7871	July 25, 1942	Retired 12/57. Sold for scrap 2/62
727	7872	July 29, 1942	Retired 8/60. Sold for scrap 11/61
728	7873	July 30, 1942	Retired 9/57. Sold for scrap 4/61
729	7874	July 30, 1942	Retired 10/57. Sold for scrap 4/61

Lima Locomotive Works, Construction Order 1166, March-May 1943.

Number	Builder's Number	Delivered	Disposition
730	8003	March 18, 1943	Retired 8/55. Sold for scrap 11/61
731	8004	March 18, 1943	Retired 11/57. Sold for scrap 2/62
732	8005	March 20, 1943	Retired 8/56. Sold for scrap 3/62
733	8006	March 22, 1943	Retired 5/57. Sold for scrap 2/62
734	8007	March 27, 1943	Retired 1/58. Sold for scrap 1/62
735	8008	April 5, 1943	Retired 8/60. Sold for scrap 2/62
736	8009	April 12, 1943	Retired 8/57. Sold for scrap 2/62
737	8010	May 21, 1943	Retired 8/60. Sold for scrap 2/62
738	8011	May 24, 1943	Retired 8/60. Sold for scrap 2/62
739	8012	May 29, 1943	Retired 2/58. Sold for scrap 5/61

A pair of S-1 Berkshires led by 716 heads out on a Wheeling District mine run to collect loaded coal hoppers. After Wheeling's four I-3 class 2-6-6-2 Baldwins and six ex-Norfolk & Western J-1 class Mountains were retired during the early 1950s, displaced NKP Berks like these two were reassigned to the Ohio coal fields to help fill the power void.
(NKP Photo by Willis A. McCaleb; John B. Corns Collection)

With a caboose tucked in behind the tender, 727 works a Conneaut-bound wayfreight west through Dunkirk, New York, at 12:20 pm on April 19, 1958. This Berkshire was retired just over two months later, on June 23.
(Fred Sankoff; Paterson-George Collection)

THE S-2 CLASS:

Chapter 4

S-2 number 753 retains its as-delivered appearance after a shopping at Conneaut in July 1946. Tender lettering includes the short-lived "swinging R" variant in the word "Road," which was applied to a variety of Nickel Plate equipment between June 1937 and February 5, 1947.
(NKP Photo by George J. Morris; John B. Corns Collection)

Even with the arrival of the S-1 class in 1942 and 1943, NKP did not use its Berkshires east of Cleveland, instead relying on aging Mikados to forward the manifests handed over by 2-8-4s at Bellevue. The reason was simple, but daunting – the steel viaducts built to carry the Nickel Plate's Cleveland and Buffalo Divisions over the many deep ravines and creek valleys flowing into Lake Erie could not support the weight of a Berkshire. This was the same obstacle that had forced the roundabout delivery of the Road's original Alco 2-8-4s in 1934. The solution had been obvious for years, but only with the unprecedented burden of wartime traffic could the comprehensive rebuilding and strengthening of these bridges be justified. The conundrum,

More War Babies

of course, was that the bridges had to be strengthened without disrupting the increased traffic in any appreciable way.

The first Nickel Plate S-2 was delivered to the Road's South Lima yard on January 11, 1944, eight months after construction of engines 740-754 had been authorized by the War Production Board. All fifteen were on the property by March 8, 1944. A follow-up order for fifteen more S-2 engines was delivered in August and September of 1944. An S-2 (without tender) was 11,900 pounds heavier than an original Alco Berkshire, and the new engines differed most notably from earlier classes in having cast frames with integral cylinders, and – 13 years after the *Four Aces* tested on NKP rails – Timken roller bearings applied to their drivers and pilot truck. They sported an additional pair of sanders per side, with the rearmost pair of drivers fitted with sand pipes fore and aft. The factor of adhesion for these engines differed slightly from the earlier Berkshires', at 4.12, while the boiler-to-cylinder horsepower ratio remained at 114.9 percent. All S-2 Berkshires were outfitted with ATS equipment.

The second group of S-2 locomotives had to forego bronze bells in favor of acoustically impure steel to support the war effort (many of the steel bells were painted yellow, though). Interestingly, Lima reverted to brass for the S-2 builder's plates after using cast iron for the S-1 locomotives.

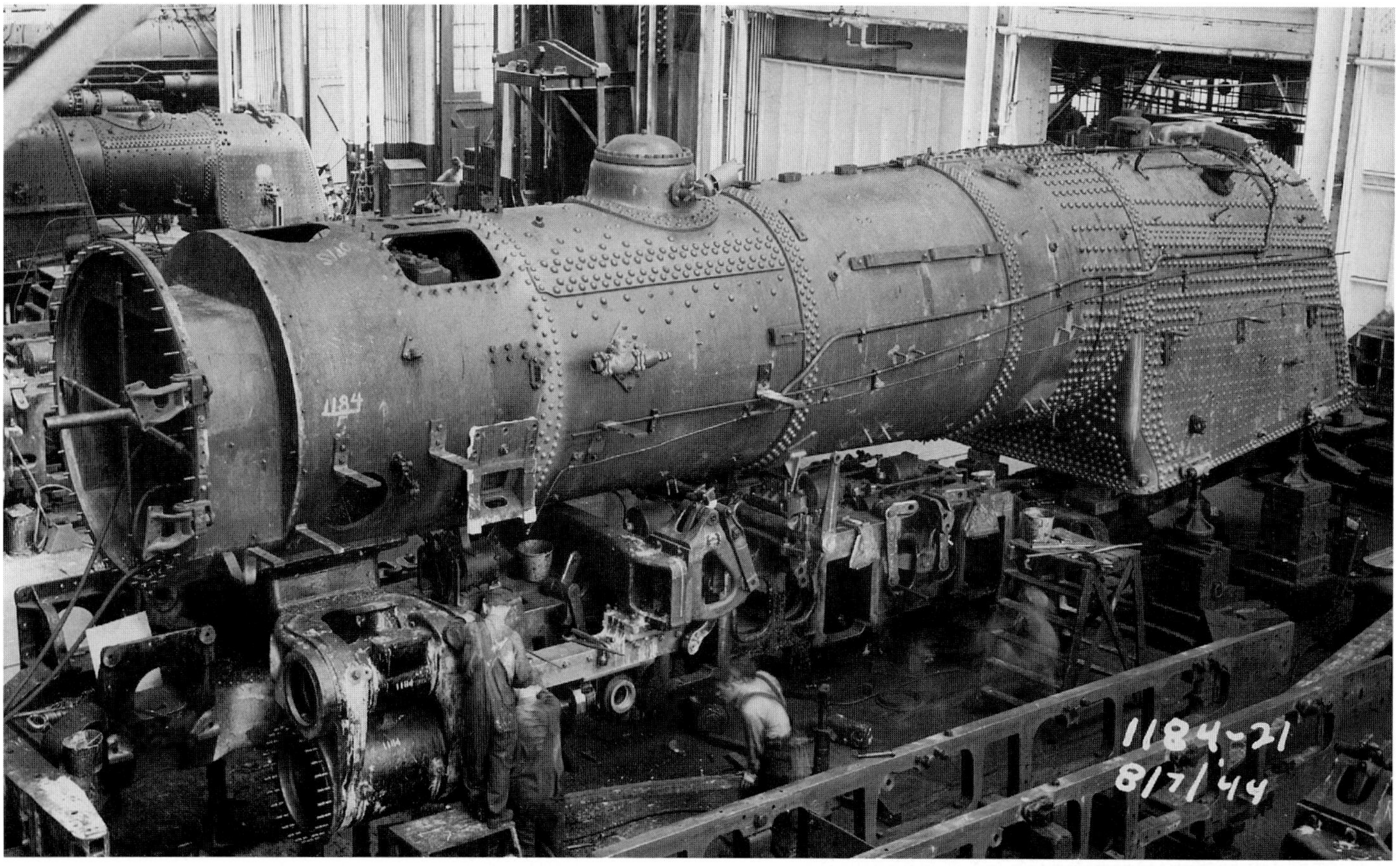

(top) Where Alco employed a flat car to trundle boiler and firebox assemblies to the erecting shop (page 19), Lima preferred to use shop trucks. In due course, this ersatz 4-4-4 will become Nickel Plate S-2 number 766.

(above) The future 759's boiler and firebox have been mated to the Commonwealth cast bed inside Lima's erecting hall, as workers attend to the left-hand cylinders and crosshead guide. The completed locomotive was delivered just 18 days after this photo was taken by Lima photographer Don Jardine.

(2 Lima Locomotive Works Photos; John B. Corns Collection)

S-2 number 742 was the third member of the class, part of Lima construction order 1179. The locomotive's handsome front end maintained the AMC features of the original S class (page 21), and is shown to good advantage in this spring 1944 portrait at Avery, Ohio.

(NKP Photo by William Rittase; John B. Corns Collection)

The final locomotive built as part of Lima construction order 1179 (and the last delivered without "flying" numberboards) has six Sherman M-4 battle tanks behind the tender in this 1944 view. The Lima Locomotive Works produced approximately 4,000 of these tanks during the war, employing the same casting and fabricating expertise used to build the "War Baby" Berkshires.
(Jay Williams Collection)

During the summer and fall of 1943, the Nickel Plate's Bridge & Building crews upgraded the viaducts east of Cleveland, clearing the way for assignment of the S-2 class to the Cleveland and Buffalo Divisions. Although they would be prohibited from reaching Peoria until a 1947 bridge upgrade at Mackinaw Dells, Illinois, the Berkshires now roamed the Nickel Plate and Clover Leaf District main lines (thanks to similar upgrading between Frankfort and Madison). For a short time, some Berkshires even ventured into passenger service.

Eight 4-6-4 Hudsons had been responsible for the Nickel Plate's modest mainline passenger schedules after 1927, sharing duties with Pacifics and obsolete Ten-Wheelers on the Road's few locals. After Pearl Harbor, the Hudsons were overtaxed by heavy consists and troop trains, to the degree that they eventually had to be rebuilt as a stopgap until NKP's passenger services could be dieselized in 1947-48. Since the Berkshires had the double blessing of being able to start heavy consists and maintain fast schedules, a total of 23 engines from the S-1 and S-2 classes were equipped with steam and communication lines and pressed into passenger operation during the latter part of the war (see Table 4).

The third week of October 1945 provides a snapshot of the Berkshires' versatility in meeting the demands of demobilization. The Nickel Plate operated troop trains either as extras or as following sections of its few scheduled passenger trains. Additionally, all military movements were assigned a "MAIN" number by the War Department (for Military Authority Identification Number). On October 23, S-1 number 716 led the 14 cars and 716 passengers of MAIN 40138/40139 (as NKP Extra 716 West) from Bellevue to Lima, where six cars and 167 riders were set off for a B&O connection before the train continued west to Frankfort. Four

(above) With the war's end in sight, Nickel Plate train 7 (the *Westerner*) pauses at Fostoria, Ohio, just past noon on July 2, 1945, behind the 766.
(Richard J. Cook, Sr.; Allen County Historical Society Collection)

(right) The *Nickel Plate Limited* (train 6) running east of Willoughby, Ohio, behind the 769 on October 26, 1946.
(Lou Sabetto)

days later, no fewer than three MAINs crossed the old LE&W route from Fostoria's FS Tower to Lima and Frankfort. Extra 715 West (MAIN 40774) on October 27 had 22 cars and 623 passenger. Later that day, Extra 719 West (MAIN 41073) had 13 cars and 316 passengers, and S-1 number 718 led a westbound extra (MAIN 40121) with 12 cars and 322 homeward-bound troops.

Some Berkshires retained their steam-heat appliances and were assigned to troop movements during the Korean War, and even led a few NKP-sanctioned excursions in the late 1950s. The 767 gained local fame when it led the first official passenger train over Fort Wayne's main line grade separation project in October 1955. Quite prophetically, S-2 number 765 became the last Berkshire (and the last NKP steam locomotive) to power a Nickel Plate passenger train when it headed a Fort Wayne-Chicago farewell excursion on June 11, 1958.

Table 4
NKP Berkshires Equipped For Passenger Service, 1944-1946

Number	Class	Builder	Built
715-719	S-1	Lima	6/42
722	S-1	Lima	6/42
729	S-1	Lima	7/42
730, 731	S-1	Lima	3/43
733, 734	S-1	Lima	3/43
735	S-1	Lima	4/43
739	S-1	Lima	5/43
743, 744	S-2	Lima	1/44
762	S-2	Lima	8/44
763-769	S-2	Lima	9/44

Source: NKP Mechanical Records.

757
757
202

763
763
763

(left) On May 19, 1946, the 757 led its freight consist past one of the 22 Cleveland Union Terminal P-1a electrics at East 37th Street. Until the Road's eleven PA-1s arrived in 1947-48, these CUT motors handled NKP passenger trains for 3.1 miles since steam power was not permitted inside CUT's station tracks. Nickel Plate steam locomotives assigned to operate in "electrified territory" (in the vicinity of CUT and at the Chicago, South Shore & South Bend interchange at Michigan City, Indiana) had their wire-reinforced whistle cords replaced with non-conductive rope cords, as a precaution against the wire from a snapped whistle cord flailing upward against the live catenary and placing crew members at risk of electrocution.

(facing page, bottom) A year earlier, on May 24, 1945, S-2 number 763 led an extra out of the East 37th Street yard. This locomotive earned a reputation as one of the best-running Berkshires on the Nickel Plate, and survives on static display in the Virginia Museum of Transportation at Roanoke.
(Both, Richard J. Cook, Sr.; Allen County Historical Society Collection)

(below) This post-1950 view recorded the 763 leading a westbound manifest across Cleveland's Cuyahoga River valley.
(NKP Photo by Willis A. McCaleb; John B. Corns Collection)

(above) Berkshires 768 and 750 lead a westbound extra of empty reefers out of Bellevue yard on July 23, 1949. One of the S-2 pair will be dropped at Fort Wayne to balance power.

(Richard J. Cook, Sr.; Allen County Historical Society Collection)

(below) The 747 makes easy work of a westbound manifest between Avery and Shinrock, Ohio, in 1948.

(NKP Photo by John D. Burger; Jay Williams Collection)

(facing page) S-2 number 749 has taken on its final appearance in this mid-1950s view, having had flying numberboards applied in 1947 and a Mars light in 1950.

(John A. Rehor; John B. Corns Collection)

749

The 744 confronts two different modes of competition in this pair of scenes from late 1957. The S-2 leads a westbound through the swing bridge gauntlet track at Lorain, Ohio, and passes a Great Lakes freighter undergoing repair.

(John A. Rehor; John B. Corns Collection)

(below) At Westfield, New York, the Berkshire puts on a show of strength as it hustles NCS-5 past a laggard New York Central westbound. Plume of steam below firebox is exhaust from the S-2's centrifugal cold water pump.

(Jim Shaughnessy)

(above) Nickel Plate rebuilt its movable span over the Cuyahoga River in 1957, and in this view 744 and an unidentified S-1 are about to pass through the new lift bridge and then under a highway viaduct still under construction. Compare with page 53.

(below) One last view of 744, at rest in the Bellevue engine terminal on July 18, 1957.

(Both, John B. Corns Collection)

740
740
740

(facing page) The 740's exhaust obscures trailing 736 as a *Flying Saucer* explodes from beneath the Ohio Turnpike at Avery, Ohio, on Labor Day weekend, 1957.

(below) Three S-2 Berkshires meet near Bellevue in 1957.
(Both, Don Wood)

(above) S-2 756 has hotshot CB-12 in tow as it parallels the Maumee River through Ft. Wayne, en route to Bellevue and Buffalo from Chicago.
(NKP Photo by Willis A. McCaleb; John B. Corns Collection)

763
763
763

753

Number 763 was regarded by crews as one of the best NKP Berkshires. Here, the S-2 leads a westbound train through Athol Springs, New York, at 11:35 am on October 26, 1952.
(J. F. Beveridge; Paterson-George Collection)

(left) Once again, form follows function. The 753 and 767 represent both construction orders of the S-2 class as they simmer at East Wayne in 1955. Their pilot-to-pilot coupling affords an opportunity to compare the S-2s' left and right sides forward of the steam dome. Of particular note are the different headlight mounts. The 753 has the original shelf, which was bolted to the smokebox door and provided a bracket for mounting the engine's rectangular number plate. On the other hand, the 767 has been retrofitted with a shelfless mount, necessitating the "gabled" style of number plate with a concave top that attached directly to the underside of the Pyle headlight case. Both engines' SBWE-1000 Mars white signal lights rest on their standard shelves above the headlight.

(below) Original numberboard style, at left, and "gable" replacement (with rare painted numerals) used with shelfless Pyle aluminum headlights. Cast metal numerals were painted yellow, except for those applied at Frankfort which were silver.
(3 Photos, John B. Corns Collection)

(right) The tail end of Chicago-Buffalo manifest OB-2 is still within Conneaut's yard limits as S-2 number 743 accelerates eastward, simultaneously crossing over the Bessemer & Lake Erie and Conneaut Creek in this 1957 scene.
(Jim Shaughnessy)

Roster of NKP S-2 Class Berkshires

Lima Locomotive Works, Construction Order 1179, January-March 1944.

Number	Builder's Number	Delivered	Disposition
740	8414	January 11, 1944	Retired 8/60. Sold for scrap 8/63
741	8415	January 14, 1944	Retired 8/60. Sold for scrap 2/62
742	8416	January 20, 1944	Retired 8/60. Sold for scrap 8/63
743	8417	January 24, 1944	Retired 8/60. Sold for scrap 2/62
744	8418	January 26, 1944	Retired 6/58. Sold for scrap 10/61
745	8419	January 31, 1944	Retired 8/60. Sold for scrap 10/61
746	8420	February 6, 1944	Retired 8/60. Sold for scrap 11/63
747	8421	February 11, 1944	Retired 8/60. Sold for scrap 12/63
748	8422	February 16, 1944	Retired 8/60. Sold for scrap 10/61
749	8423	February 18, 1944	Retired 8/60. Sold for scrap 12/63
750	8424	February 26, 1944	Retired 8/60. Sold for scrap 1/64
751	8425	February 28, 1944	Retired 8/60. Scrapped by NKP 10/60
752	8426	March 1, 1944	Retired 8/60. Sold for scrap 11/63
753	8427	March 6, 1944	Retired 6/58. Sold for scrap 4/61
754	8428	March 8, 1944	Retired 8/60. Scrapped by NKP 10/60

Lima Locomotive Works, Construction Order 1184, August-September 1944.

Number	Builder's Number	Delivered	Disposition
755	8663	August 14, 1944	Retired 8/60. Preserved at Conneaut 10/64
756	8664	August 16, 1944	Retired 8/60. Sold for scrap 10/63
757	8665	August 22, 1944	Retired 8/60. Preserved at Strasburg, Pa., 1966
758	8666	August 23, 1944	Retired 2/58. Sold for scrap 4/61
759	8667	August 25, 1944	Retired 8/60. Sold to F. Nelson Blount 10/62
760	8668	August 26, 1944	Retired 1/58. Sold for scrap 4/61
761	8669	August 28, 1944	Retired 8/60. Sold for scrap 10/63
762	8670	August 30, 1944	Retired 8/60. Sold for scrap 11/63
763	8671	Sept. 1, 1944	Retired 8/60. Preserved at Roanoke, Va., 7/66
764	8672	Sept. 5, 1944	Retired 6/58. Sold for scrap 10/61
765	8673	Sept. 8, 1944	Retired 8/60. Preserved at Fort Wayne, 5/63
766	8674	Sept. 10, 1944	Retired 5/58. Sold for scrap 10/63
767	8675	Sept. 14, 1944	Retired 4/58. Sold for scrap 10/63
768	8676	Sept. 22, 1944	Retired 6/58. Sold for scrap 1/62
769	8677	Sept. 28, 1944	Retired 6/58. Sold for scrap 1/62

(facing page) The next-to-last S-2 hustles a reefer train through Old Fort, Ohio, (between Bellevue and Fostoria) on March 21, 1957.
(John B. Corns Collection)

(above) Nickel Plate was an early proponent of piggyback – the hauling of highway trailers on flatcars – and accordingly was one of the few railroads that hauled trailers behind steam. In this view, the 748 has a half-dozen piggyback cars in tow as it heads east out of Bellevue at 7:30 am on August 23, 1955.
(J. F. Beveridge; Paterson-George Collection)

There's more than 22 tons of coal in 769's tender as the youngest member of the Nickel Plate's S-2 class awaits its next assignment at Chicago's Calumet terminal on July 13, 1955.
(J. R. Quinn Collection)

THE S-3 CLASS:

Chapter 5

Competitive motive power philosophies meet face-to-face as NKP Extra 778 West whips through the S-curve at Silver Creek, New York, past a detouring NYC eastbound behind F7A 1808.
(Jim Shaughnessy)

By 1947, the fate of the reciprocating steam locomotive in the United States seemed to be sealed, particularly from the builders' perspective. Two of the Big Three – Alco and Baldwin – were making the wholesale transition from steam to diesel production, attempting to claim their shares of the internal combustion market not already dominated by General Motors (by virtue of the FT's prewar success and GM's War Production Board-sanctioned head start in road locomotive production).

The Lima Locomotive Works acknowledged the inevitable that year by merging with the General Machinery Corporation of Hamilton, Ohio. In doing so, the company became Lima-Hamilton and was assured a secure in-

END OF AN ERA

house supply of diesel prime movers, although its first diesel-electric switcher would not roll out of the Works until the spring of 1949.

From most railroads' perspective, the outlook was much the same – full dieselization was either already accomplished or was an eventuality. There were a few notable exceptions – coal haulers like C&O, Norfolk & Western, and Louisville & Nashville – which were not only committed to maintaining steam operations as the 1950s approached, but were convinced that the purchase (and, in N&W's case, construction) of new steam locomotives made economic and political sense. The Nickel Plate counted itself among the exceptions, even as the Road put its Berkshires to the test against General Motors' latest road freight diesel offering.

In June 1948, EMD sent an A-B-B-A set of F3s (borrowed from the Chicago, Burlington & Quincy, numbers 116A-116D) to make two round-trip test runs over the NKP's fast, flat Nickel Plate District (Chicago-Buffalo). The tests commenced in Chicago on June 15 (with train OB-2) and ended there on the morning of June 20 with the arrival of train BC-1. At that time, the only diesels in NKP service were eleven brand-new Alco PA1s on the Road's mainline passenger trains, and 43 EMD, Alco, and Baldwin switchers assigned to Chicago, Buffalo, Cleveland, and smaller outlying terminals. The Nickel Plate was shopping for new road locomotives to replace

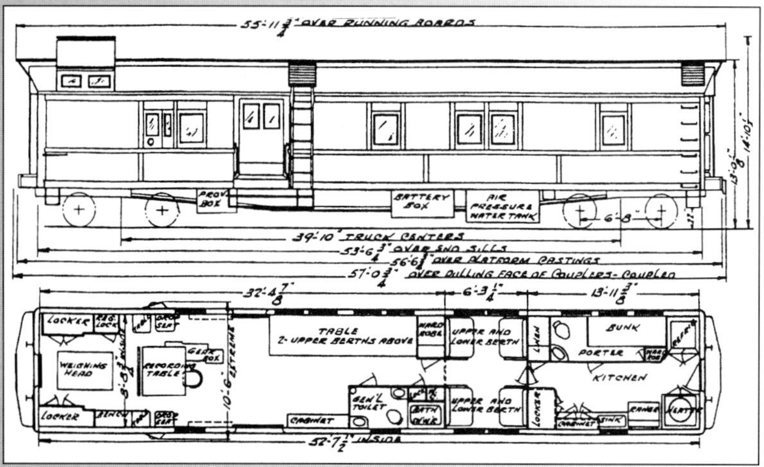

(top) Dynamometer Car X-50041 played a pivotal role in the Nickel Plate's development. Built by Standard Steel Car as Lot 113 in May 1924, the car could accommodate a test crew of six and a porter. In this photo, NKP test engineer Don Coon reviews the Buffalo Division profile during March 1949 gas turbine tests held in conjunction with General Electric.

(GE Photo; TLC Collection)

(middle) NKP X-50041 is shown at Bellevue in 1965. Following its 40-year NKP career, the car went on to serve Alco and the U.S. Department of Transportation before being preserved in Bellevue.

(Howard W. Ameling)

its aging Mikados, and was also studying the costs related to dieselizing four key Chicago-Buffalo freight schedules. For its part, EMD intended to see that its F-units filled the order. Unbeknownst to the executives at General Motors, however, a decision had already been made.

The F3, as its model designation suggested, was the third design evolution of the road freight locomotive introduced by Electro-Motive as the FT in 1939. With the F3, 1500-hp cab or booster units were intended to be combined building-block fashion into 3000-hp, 4500-hp, or 6000-hp "locomotives" on the road – an approach derived from the earlier FT and F2, but with units of greater horsepower and improved component design.

With NKP's well-traveled dynamometer car X-50041 measuring their performance, the F3s were tested in 6000-hp and 4500-hp combinations, and evaluated against previously documented S-2 Berkshire speed and horsepower results. The F3s stayed together as a four-unit "locomotive" during the test runs, with one unit towed dead during the 4500-hp trips. These test runs were not head-to-head comparisons of steam versus diesel, although two Chicago-Buffalo round trips were made to compute steam locomotive fuel costs, in which Berkshires hauled "the same approximate tonnage and on same [sic] schedules" as two of the documented F3 trips.

Below about 30 miles per hour, diesel outperformed steam in terms of both drawbar horsepower and speed. At the higher speeds required by NKP's freight schedules, the three-unit (4500-hp) F3 produced only about 95 percent of a Berkshire's average actual drawbar horsepower – at a fractionally higher fuel cost – without being able to match an S-2's time over the District. At least that was what the "official" statistics indicated (see Table 5). Below 30 mph, the 6000-hp F3 yielded an average drawbar horsepower of 4,699 as opposed to 2,130 for an S-2. Above 30 mph, the F3 registered an average of 5,001 against a Berkshire's 3,915. The four-unit A-B-B-A (6000-hp) F3 outperformed the steam power through all speed ranges, but this was reflected in fuel costs that were almost 16 percent higher than the Berkshires' – again, according to "official" numbers

Table 5
S-2 Berkshire vs. EMD F3
Chicago-Buffalo Official Test Results, June 15-20, 1948.

	S-2 Berkshire	F3 (4500-hp)
Drawbar hp <30mph	2,130	3,338
Drawbar hp >30mph	3,915	3,750
Fuel cost per 1,000 gross ton-miles	17.95¢	17.96¢
Net transit time (508 miles)	12:45	13:30

Source: Railway Progress, December 1953, p.28.

which were, at best, selective. The record is clarified by a four-page letter, dated July 16, 1948, from NKP Chief Mechanical Officer T.C. Shortt and General Manager A. R. "Gus" Ayers to Vice-President of Operations F. S. Hales, in which they summarized the F3 test results. They noted that the 4500-hp F3's duly publicized 13-hour, 30-minute Chicago-Buffalo transit time "was adversely affected by a broken rail and other operating conditions on the Fort Wayne Division, which necessitated the train being stopped four times and consequently having the net time materially increased." In total, the 4500-hp F3 test train made ten stops en route, versus only three for the comparable Berkshire-powered trip. "For purpose of net time comparison," Shortt and Ayers continued, "it is considered equitable to eliminate this division's performance both in steam and diesel operations from this report. By making this elimination, the aggregate net time performance on the other three divisions is tabulated as follows – eastbound net time under steam power: nine hours, 45 minutes; eastbound net time under 4500-hp [F3] locos: nine hours, 48 minutes." They concluded that the diesel would actually hold a time advantage both eastbound and westbound, since it had still made more stops than the Berkshire even after the Fort Wayne Division's performance had been discounted. They went on to report that the 6000-hp F3 ran Chicago-Buffalo in 12 hours and eleven minutes, including nine stops. This was 34 minutes faster than the steam comparison, which only made three stops. Even with five more stops than the steam comparison, the westbound 6000-hp F3 test train completed its run one hour and eight minutes faster.

The Shortt-Ayers letter also reveals that the "official" fuel costs for steam operation, while incorporating consumption for such "terminal standby operations" as fire cleaning, fire dumping, and boiler blowdown, were for coal only. The Berkshire fuel statistics released to the industry did not include costs for treated water, which Shortt and Ayers calculated to be 17.94¢ per thousand gallons. Had the cost of water been included in the 4500-hp comparisons, the Berkshires' true fuel costs (see Table 5) would have risen to 18.63¢ per thousand gross ton-miles.

Shortt and Ayers stopped shy of making any recommendations in their letter, but concluded that "the total expenses to dieselize the [four] trains under consideration by the use of the 4500-hp [F3] locomotive would increase $73,093.00 yearly as compared with the operation of the above-mentioned trains with the S-type steam power." They determined that the extra costs associated with adopting 6000-hp F3s for the four trains would amount to $266,613.00 annually, with the difference accruing from "higher fixed charges against equipment and facility expenditures required for the 6000-hp locomotive," as well as proportionally greater maintenance expenses.

On July 2, 1948, not even two weeks after the F3s were sent back to La Grange, NKP surprised the industry and ordered ten Berkshires from Lima-Hamilton. This $2.25 million deal was to prove one of only two instances in which an American railroad re-ordered steam freight power after assessing diesels for the same service (the second, not surprisingly, was Norfolk & Western). The decision, according to Lynne White (who had become NKP president upon John Davin's death in 1949), had been based exclusively on assessment of the F3 test data, despite later industry speculation that the results had been manipulated by NKP in favor of a sixth Berkshire order. When asked in 1953 about the decision to purchase the S-3 class, White observed, "First, for our kind of operation – trains of moderate length running over flat country – we don't require the great pulling power of the diesel. Second, our present steam power is relatively new, is designed for our kind of operations, and therefore, there is little difference in operating costs. Needless to say, we ordered the Berkshires."

(continued on page 72)

How To Build A Berkshire

(Lima Locomotive Works; John B. Corns Collection)

One of the five Nickel Plate Road Locomotive Inspectors assigned to follow Lima order 1209 through the various stages of assembly, H. D. Miller had a front-row seat during construction of these ten locomotives. His April 1949 report detailing this process provides a remarkable glimpse at the end of an era, and it is reproduced here through the courtesy of John H. Keller and Charles Bates of the Allen County (Ohio) Historical Society.

"On July 2, 1948, the Nickel Plate Road placed the order for the Class S-3 locomotive with the Lima-Hamilton Corporation. It took the Lima concern approximately six months to order and produce the material to make up the 40,000-odd parts that go into a Berkshire locomotive. During the last few months of 1948, the ordered parts began to flow into Lima from all parts of the United States.

"The actual work on the new freight haulers commenced on the second of January. As the work began, five Nickel Plate Motive Power Inspectors were assigned to Lima to inspect the new locomotives as they were being built. These Frankfort and Conneaut inspectors worked along with the Lima-Hamilton Corporation's Inspection Department to insure that the locomotives were built in the best of condition. The work of building a locomotive at Lima is divided between the erecting floor, the boiler shop, the tank shop, the paint shop, and the truck shop.

"In the boiler shop, the various sheets of boiler plate were laid out and drilled. These sheets vary in thickness from 9/16 inch for the smoke box to 1-1/16 inch for the third course. After the rivet holes are drilled, the flat plates are rolled or flanged into cylindrical and firebox shapes. After the rolling process, the seams are welded together and welt straps are riveted in place. The three boiler courses, smoke box, and the firebox are then riveted together by a giant riveting machine called the Bull. The dry pipe and boiler braces are applied while the studs are being pulled in. The next step in the boiler construction is the application of the 1,832 staybolts to the firebox. Next, the 202 flues and 73 tubes are applied. All of this boiler work is thoroughly checked by the Nickel Plate Inspectors to insure that the boiler is perfectly leak-proof.

"The boiler is then given a water test. It is filled with hot water and the pressure is then raised gradually to 25% above the designed pressure of 250 pounds per square inch. All leaks are repaired during this hydrostatic test. The next test to be applied is one with live steam. The pressure is gradually raised to 250 pounds per square inch over a three-and-one-half hour period. All leaks that show up under this test are repaired. The boiler is then allowed to cool down and it is thoroughly washed and the same test is again applied. If the boiler is found leak-proof by the Nickel Plate Inspectors it is painted with red lead and made available to the erecting floor. The complete boiler, freshly painted, weighs 103,500 pounds.

"While the boiler is being manufactured, the thirty-five-ton locomotive bed frame with cylinders cast integral is placed on the erecting floor. The valve and cylinder bushings are pressed into the bed frame and then the cylinder is bored to 25 inches and the valve to 14 inches. The Baker valve gear frames, reverse shaft, brake cylinders, spring rigging, chassis lubrication lines, and the pilot are applied to the frame.

"At this stage, the completed boiler is placed on the frame and bolted down. The superheater units are placed in the flues while the under-the-jacket piping is applied to the boiler. The boiler is then well insulated with standard Magnesia block lagging of at least 1-1/2 inch thickness. A jacket of cold rolled steel is then fastened over the lagging. The locomotive is now ready to be wheeled. The four sets of driving wheels and the engine truck wheels are mounted on axles supported in roller bearing housings. The trailer truck and engine trucks are assembled in the truck shop. The cab, which has been fabricated in the cab shop, is then mounted on the locomotive. The cab on the S-3 locomotive is new to the Nickel Plate in that the cab walls and ceiling are insulated with Insulmat. This is the same material that is sprayed underneath new automobiles. Its advantages are that it deadens noise, prevents condensation, and insulates the cab against heat and cold.

"Temporary wooden main rods are used to facilitate the setting of the Baker valve gear. The main driving wheels are set up on motor-driven rollers during the valve setting. Next, the pipefitters are at work installing piping around the engine and in the cab. The lubricating oil pumps, whistle, separators, stoker, injector, and thousands of other essential parts are installed before the engine is ready for a fire. The average time elapsed after the boiler is placed on the frame until the locomotive is ready for test is approximately nine working days.

"While the erecting floor and boiler shop work is in progress, the tank shop is busy building the standard Nickel Plate class 22-RG tenders. These tenders are capable of holding 22,000 gallons of water and 22 tons of coal. The completed tender when loaded with coal and water weighs approximately 368,000 pounds. At the Lima-Hamilton Corporation plant it takes approximately two weeks to build a Nickel Plate tender.

"The locomotive and tender are now ready for the running test. The locomotive is placed under a full head of steam and run for about four hours at speeds of ten miles per hour. During this shakedown run, the Nickel Plate and Lima-Hamilton inspectors check over the machinery and boiler to insure the locomotive will be in excellent condition when it is put on a Nickel Plate fast freight. All defects are reported and corrected. During the running test, the locomotive is taken through a clearance hole that is exactly 15 feet, 10-5/8 inches high and 11 feet wide. The locomotive must pass through this hole without scraping the cloth screen that borders the hole. The overall length is checked to make sure that it is 100 feet, 8-3/4 inches from the [front] coupler knuckle to the rear tender coupler knuckle. The locomotive is also weighed to make sure that it is not overweight in regard to the capacity of the Nickel Plate roadbed and bridges. A separate scale is placed under each wheel of the locomotive and tender. In this way, the weights on the drivers and trucks can be accurately checked to insure that the weight is well distributed. The Class S-3 freight puller weighs 812,200 pounds when she has a full tender of coal and water and when the boiler is up to the working pressure of 245 pounds per square inch.

"After the running test, the locomotive is taken to the paint shop where it is thoroughly

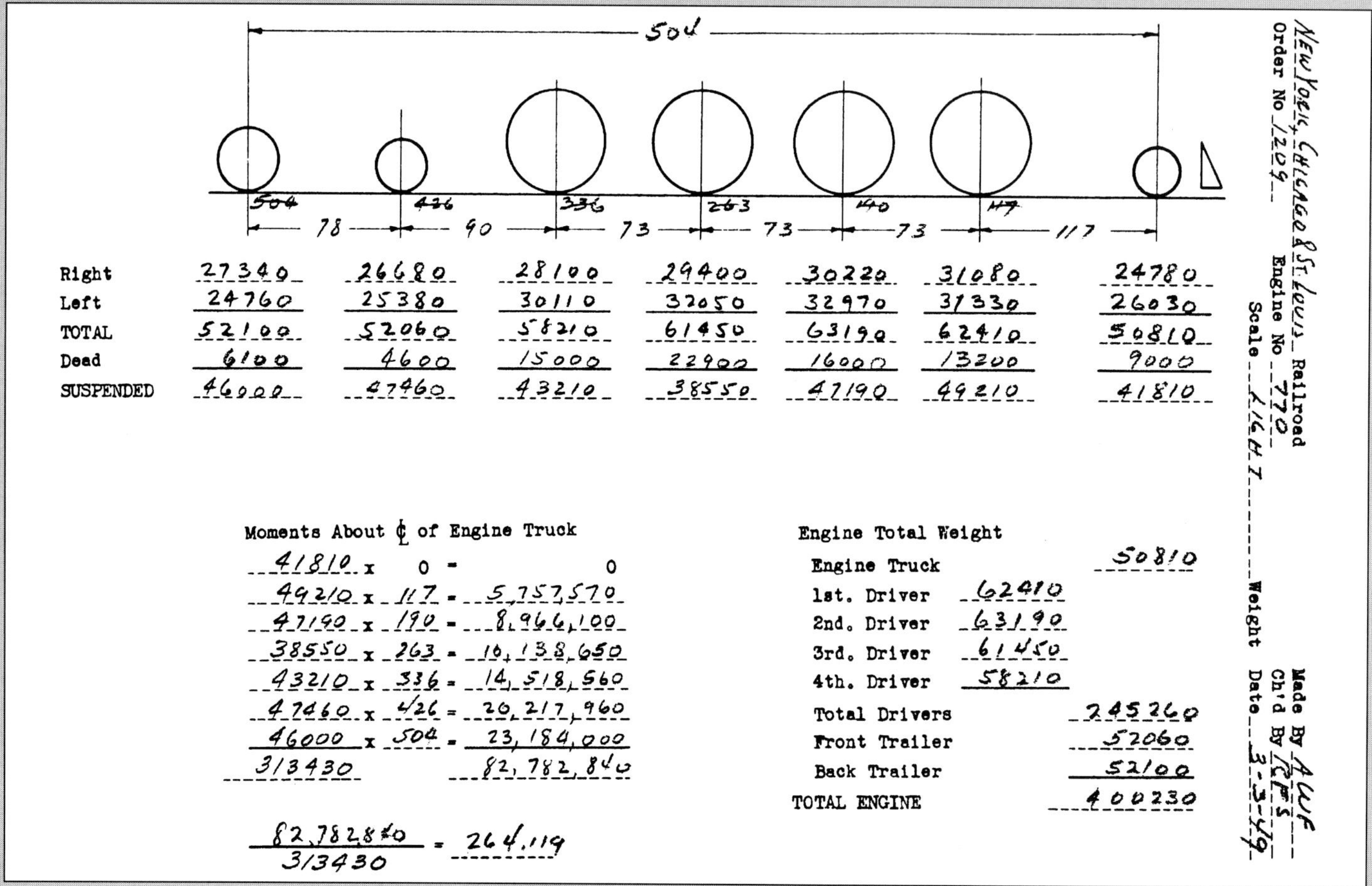

New York, Chicago & St. Louis Railroad
Order No 1209
Engine No 770
Scale LIGHT Weight
Made By AWF
Ch'd By RFS
Date 3-3-49

504

	78	90	73	73	73	117	
	504	426	336	263	190	117	
Right	27340	26680	28100	29400	30220	31080	24780
Left	24760	25380	30110	32050	32970	31330	26030
TOTAL	52100	52060	58210	61450	63190	62410	50810
Dead	6100	4600	15000	22900	16000	13200	9000
SUSPENDED	46000	47460	43210	38550	47190	49210	41810

Moments About ℄ of Engine Truck

41810	x 0	=	0
49210	x 117	=	5,757,570
47190	x 190	=	8,966,100
38550	x 263	=	10,138,650
43210	x 336	=	14,518,560
47460	x 426	=	20,217,960
46000	x 504	=	23,184,000
313430			82,782,840

$$\frac{82{,}782{,}840}{313430} = 264.119$$

Engine Total Weight

Engine Truck		50810
1st. Driver	62410	
2nd. Driver	63190	
3rd. Driver	61450	
4th. Driver	58210	
Total Drivers		245260
Front Trailer		52060
Back Trailer		52100
TOTAL ENGINE		400230

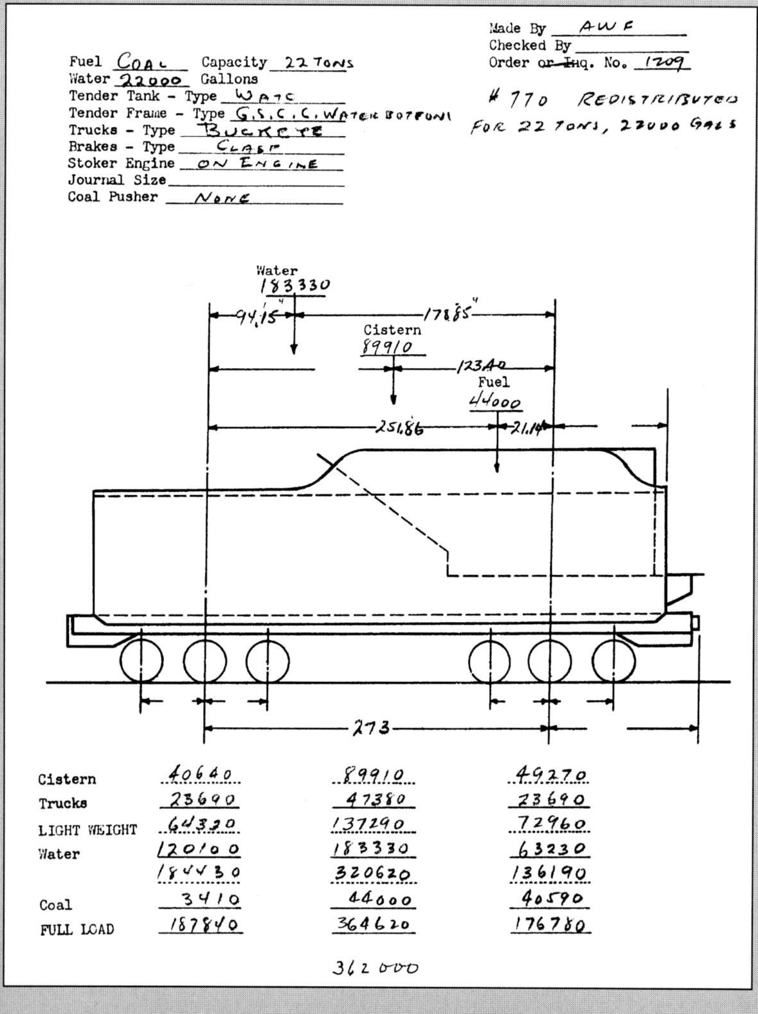

Made By AWF
Checked By
Order or Inq. No. 1209

#770 REDISTRIBUTED FOR 22 TONS, 22000 GALS

Fuel COAL Capacity 22 TONS
Water 22000 Gallons
Tender Tank - Type WATC
Tender Frame - Type G.S.C.C. WATER BOTTOM
Trucks - Type BUCKEYE
Brakes - Type CLASP
Stoker Engine ON ENGINE
Journal Size
Coal Pusher NONE

Cistern	40640	89910	49270
Trucks	23690	47380	23690
LIGHT WEIGHT	64330	137290	72960
Water	120100	183330	63230
	184430	320620	136190
Coal	3410	44000	40590
FULL LOAD	187840	364620	176780

362000

The calculation – and recalculation – of weights and weight distribution was a critical and dynamic element in steam locomotive construction, even in 1949. Despite early efforts to keep the S-3 Berkshires within the same total operating weight as the previous S-2 class, they ended up being over three tons heavier. The two weight calculations on this page were produced by Lima-Hamilton engineers during construction of S-3 number 770. The weights at the top of the page are for the light engine, while those at left are for the fully loaded tender. A locomotive's exact weight was not known until construction was complete. This figure was of vital importance to the railroad, and not only for ensuring that heavy equipment did not operate over light-duty trackage or bridges. Engine crews were paid based on the weight of their assigned locomotive, so it was in the railroad's best interest (if not the crew's) to produce an "official" figure that was as low as practical. In fact, the weights of the original Alco Berkshires were reportedly understated by NKP for just this reason.

(Allen County Historical Society Collection)

(facing page, top) Typically, one completed locomotive from each order was selected to pose for official builder's portraits. NKP 776 was chosen to represent the S-3 Berkshires, and none of the other nine were photographed upon completion at the Lima works. This Don Jardine "builder's photo" of 779 is, in fact, an "officially" retouched view of 776 – the numeral "6" was simply inverted to create a "9" on a print and rephotographed. By late 1949, when it had become clear that the 779 was Lima-Hamilton's final steam locomotive, requests flooded in to "The Loco" for photos of the now-famous Berkshire. Since none had been taken before delivery, this view was fabricated to satisfy the demand.

(Lima-Hamilton Photo; John B. Corns Collection)

(facing page, bottom) For steam locomotives, the inspection process was ongoing, from construction right through retirement. Monthly, quarterly, and annual inspections were mandated by the government, with the first for a given locomotive occurring upon delivery. This is 770's original inspection card.

(Allen County Historical Society Collection)

cleaned and made ready for painting. One coat of a special front-end paint is applied to the smoke box and smoke stack. The remaining parts of the locomotive are painted with two coats of [black and lettered with] imitation gold letter paint. As on all other Nickel Plate steam locomotives, the faces of the tires on the driving wheels and the edges of the running board are painted a steel color.

"After leaving the paint shop, the locomotive is again fired up and thoroughly checked from the pilot to the rear tender light by the Nickel Plate and Lima-Hamilton inspectors. When the condition of the locomotive meets with the approval of the Nickel Plate Chief Locomotive Inspector, it is taken to the Nickel Plate Lima roundhouse. There it is again serviced and made available to the Dispatcher for assignment to an eastbound through freight. The locomotive is again serviced and completely checked at Bellevue before it is sent to its new home terminal of Conneaut, Ohio. There the Class S-3 engines will earn their keep on the Buffalo and Cleveland Divisions."

I. C. C. Form No. 3 **ANNUAL LOCOMOTIVE INSPECTION AND REPAIR REPORT** Form L-128 CNP 6-39

MAR 14 1949, 19__ 1209 - 1 - 9371 Locomotive { Number 770; Initial N.Y.C. & St. L. }

THE NEW YORK, CHICAGO & ST. LOUIS RAILROAD COMPANY

In accordance with the act of Congress approved February 17, 1911, as amended March 4, 1915, and the rules and instructions issued in pursuance thereof and approved by the Interstate Commerce Commission, all parts of Locomotive No. 770, including the boiler and its appurtenances, were inspected on MAR 14 1949 19__, at Lima, Ohio and all defects disclosed by said inspection have been repaired, except as noted on the back of this report.

1. Date of previous hydrostatic test New Boiler, 1949
2. Date of previous removal of caps from flexible staybolts, New Boiler, 1949
3. Date of previous removal of flues New Boiler, 1949
4. Date of previous removal of all lagging New Boiler, 1949
5. Hydrostatic test pressure of 313 pounds was applied.
6. Were caps removed from all flexible stay bolts? 23B. Buff.
6a. Were all stay bolts hammer tested under hydrostatic pressure? Yes
7. Were all flues removed New Number removed None
8. Condition of interior of barrel Good.
9. Was all lagging removed? New Boiler, 1949
10. Condition of exterior of barrel Good
11. Was boiler entered and inspected? Yes.
12. Was boiler washed? Water glass cocks and gauge cocks cleaned? Yes. Yes.
Water column connection Yes
13. Condition of crown stays and staybolts Good. Good.
14. Condition of sling stays and crown bars Not used
15. Condition of firebox sheets and flues Good. Good.
15a. Was new firebox applied? New Boiler, 1949
16A Condition of arch tubes Good. Water bar tubes None
17. Condition of throat braces Not used.
18. Condition of back head braces Good
19. Condition of front flue sheet braces Good
20. Were fusible plugs removed and cleaned? None
20a. Give make and condition of low water alarm (Nathan type) (B. Good)
21. Were steam leaks repaired? Yes.

Items 1 to 21 and 26. I certify that the above report is correct __________, Inspector.

22. Were steam gauges tested and left in good condition? Yes
23. Safety valve set to pop at 245 pounds, 248 pounds, 251 pounds.
24B Were both injectors tested and left in good condition? Yes.
25. Were steam leaks repaired? Yes
26.C Hydrostatic test of 210 pounds applied to main reservoirs.
27. Condition of brake and signal equipment Good. Not used.
28. Were drawbar and drawbar pins removed and inspected? New
29. Condition of draft gear and draw gear Good. Good.
30. Condition of driving gear Good.
31. Condition of running gear Good.
32. Condition of tender Good

Items 12 and 22 to 32. I certify that the above report is correct __________, Inspector.

State of Ohio } ss: County of Allen }

16A Two Arch tubes, two syphons
24B One injector and one Worthington F.W. Heater
26C Builder's test

Subscribed and sworn to before me this ____ day of MAR 14 1949 19__, by { } Inspectors of the Lima Locomotive Works Div., Lima-Hamilton Corp. Company.

H. T. ROBINSON Notary Public, Allen County, O. My commission expires Sept. 15. , Notary Public

The above work has been performed and the report is approved.

__________, Officer in charge.

Table 6

Summary of S-3 Class Design Changes (vs. S-2)

(not all changes listed)

- Roller bearing truck axles to be arranged for axle end caps.
- Steel plate bell bracket for stationary mounting of bell omitting steel bracket, yoke, and trunnions. Bell to be made of bronze metal.
- Superheater units to end 18" from back tube sheet (6" shorter than previous).
- Reduce boiler steam test pressure to be same as boiler designed pressure instead of 20% in excess of boiler designed pressure.
- Reduce boiler hydrostatic test pressure to 25% above boiler designed pressure instead of 33-1/3% in excess of boiler designed pressure.
- Firebox side sheets and syphons to be molybdenum steel. Rigid staybolts in side sheets ... to be of approved analysis of molybdenum steel.
- Roller bearing housings to have bayonet-type oil level indicators.
- If roller bearing trailing truck can be used and keep within weight limitations we are to give same consideration.
- Flexible armored hose for lines between engine and tender instead of Barco joints.
- Front left side blow-off cock to discharge into muffler skirt at right side front corner of firebox.
- Watershed cab. Panel type clear vision window for front of cab. Investigate cost of Edwards metal cab side window sash. Foam rubber cab seat cushions.
- Wood lining in cab to be omitted and cab insulated with Insulmat 1/4" thick.
- Luckenheimer quick acting blower operating valve.
- Investigate lightening type of crosshead guide eliminating Brewster type wrist pin.
- Mild steel preferred over wrot iron for drawbars and safety bars.
- Eccentric crank to be lightweight design similar to C.O. 1198.
- Roller type engine truck centering device.
- Wright's little watchman valve for engine truck derailment.
- Unequal slopes of trailer truck rocker bearings for backup operation.
- Single keeper bolt in each frame shoe.
- Apply steps to trailing truck and grab handle at ash pan wings.
- Train control equipment omitted.
- Tender electric wiring in handrail.
- Valve pilot speed recorder to be used (100 mph type).
- Injector to be equipped with warning pipe. Injector operating pins to have Alemite fittings.
- Lima to submit proposal for complete chassis mechanical lubrication.
- Center hung-type driver brake shoes
- Cab heater drain pipes to be arranged to prevent freezing.
- Auxiliary reservoir for power reverse gear. Power reverse gear to be Precision type F-3.
- Back side rod to have grease cups omitted.
- Three piece welded-type driving box saddles.
- Apply emergency brake valve on tender at gangway.
- Dry pipe to be extended towards rear of boiler.
- Throttle header to be cast steel.
- Omit collar around top of tender tank of water compartment.
- Piston valves to be arranged for line and line exhaust clearance.
- Rear trailer wheels to be 44" wrot steel Edgewater.
- Three-piece stoker protecting grates.
- Louver side cab ventilator similar to C&O orders.
- Eliminate pipe connections where possible. Use welded connections.
- Weight of engine and tender cannot exceed C.O. 1184 [NKP S-2].
- Material specifications to be based on latest C&O specifications as C.O. 1205 [H-8 Allegheny 1645-1659]
- Cab arm rest arrangement to be similar to ... C&O locomotives.
- Manual operated whistle rigging to be applied as well as pneumatic whistle operator.

Sources: Lima Locomotive Works "Memorandum Of Interview," July 7, 1948; NKP internal memo, August 3, 1948.

What White didn't say was that approval for the purchase of ten S-3 Berkshires was contained in a June 17, 1948, board resolution – passed while the borrowed F3s were barely halfway through their "testing." After the S-3 order was placed with Lima on July 2, the NKP's comptroller issued Authority For Expenditure (AFE) 15854, documenting the $2,350,000 required to purchase the new Berkshires (final unit cost was $224,866.20). Dated July 16, 1948, the AFE bore the recommendation of T. C. Shortt (on the very day he and Gus Ayers had written to F. S. Hales regarding the F3 tests) and the approval of John Davin. The Nickel Plate, it turns out, never had any intention of pursuing an F3 order with EMD. The skeptics were right.

Rather than simply duplicate the five-year-old S-2 specifications, NKP and Lima took advantage of postwar materials and technology to create construction order number 1209, the S-3 class. No fewer than 97 proposed refinements over the older Berkshires were detailed in meetings between the railroad and builder during the summer of 1948. These ranged from returning to bronze bells (instead of the War Babies' acoustically undesirable steel) and deleting wood cab insulation in favor of a sprayed asphaltic compound to applying complete mechanical chassis lubrication (see Table 6).

Despite engineering directives to limit these locomotives to the same weight as the last S-2 order, the S-3 class would be the Nickel Plate's heaviest Berkshires – at 808,900 pounds total and 444,300 pounds for the engine alone, according to NKP records – with a 4.17 factor of adhesion. They shared the cast frames and roller bearings of the S-2 class, and maintained the S-2s' 64,100 pounds of tractive power and 245 psi boiler.

This May 10, 1949, portrait by Don Jardine reveals several S-3 idiosyncrasies. Polished bronze bells replaced steel versions used on the S-2 class, but on NKP 770-779 they were hung from a simple sheet metal bracket. Curiously, 770-779 had Pyle headlights with integral numberboards, even though raised boards had become standard and this more expensive Pyle fixture was therefore not necessary. As had been the case with S-2 order 1184, the S-3 class was delivered with short-lived C&O-style Futura numerals in the raised number boards. On the S-2 Berkshires, these were changed to the heavier NKP style shortly after delivery. Their presence on the S-3 engines resulted from Lima's erroneous interpretation of a July 1948 NKP directive stating "where not otherwise specified the same specialties as used on C.O. 1184 apply."
(Lima Locomotive Works; John B. Corns Collection)

776
776
776

"First of the last," the 770 heels through the east side of Cleveland with a westbound manifest. NYC/CUT tracks are at left, rapid transit in right background.
(NKP Photo by Willis A. McCaleb; John B. Corns Collection)

Some S-3 design changes were beyond the control of Lima and NKP, dictated by the reduced availability of steam locomotive components from third party suppliers forced to adapt to their own radically changing markets. One example of this occurred on July 7, 1948, when the Worthington Pump & Machinery Company's W. B. Savary advised Lima Chief Engineer M. J. Donovan that the type 5-SSA feedwater heater used on the S-1 and S-2 engines had become non-standard and would not be produced for Lima order 1209. Instead, the standard type SA – as used by the C&O – would be provided. In the conclusion of his letter, Savary observed "since that road [C&O] has the same Advisory Mechanical Committee as the Nickel Plate, we believe they will approve it for the latter road also." Even fifteen years after creating the original design, the AMC still had a hand in the Berkshires' evolution. The first S-3, NKP 770, was delivered from the Works to the Road's South Lima yard on March 14, 1949. When 779 "crossed the fence" two months later on May 13, Lima's boilermakers and other steam craftsmen unknowingly watched their livelihoods go with it. Lima-Hamilton put on a brave face for a while in the trade magazines, reaffirming its commitment to the production of steam locomotives, but its triumph over EMD had been fleeting. The 80-year-old company had built its final steam locomotive. Although the S-3 Berkshires shared construction space with Lima-Hamilton's first diesel switchers, the Works was never able to capture enough of the diesel market to remain a viable builder. A 1950 merger with Baldwin created Baldwin-Lima-Hamilton Corporation, but did little to sustain the L-H switcher line. Locomotive production at Lima ceased on September 12, 1951, with delivery of the Works' 174th diesel, PRR 2500-horsepower transfer unit 5683.

As the Road's newest Berkshire, the 779 was chosen to represent NKP at the 1949 Chicago Railroad Fair, and this engine spent most of its first four months on display by the shores of Lake Michigan. Super-Power had been instrumental in transforming the Nickel Plate, and the Road was not shy about giving credit where it was due. During the summer of 1948, almost 2.5 million people had attended celebrations marking the centennial of the

(above) The crowds have started to thin, but the 779 still holds several little (and not so little) boys enraptured at the close of the 1949 Chicago Railroad Fair's *Wheels A-Rolling* pageant. How many in attendance suspected that the shimmering Berkshire was to be the last of its kind, and the final reciprocating product of a doomed builder?

LIMA HAMILTON CORPORATION

After putting in 1-1/2 months of revenue service, the 779 was spruced up at Conneaut in preparation for display at the Fair. The engine got a fresh paint job, and the as-delivered Futura number board numerals were replaced with the correct NKP style. Most significantly, 779 became the first NKP Berkshire to receive a white Mars light above the headlight, almost a year before the appliances became standard. Interestingly, the light visible in this pre-Fair view at Cleveland's East 9th Street yard featured an 18" oscillating reflector, whereas all subsequent NKP steam applications (including a refit to 779) incorporated the Mars SBWE-1000 model. This later mechanism used a standard 7-1/2" sealed beam bulb mounted in an oscillating frame.

(2 NKP Photos by John D. Burger; John B. Corns Collection)

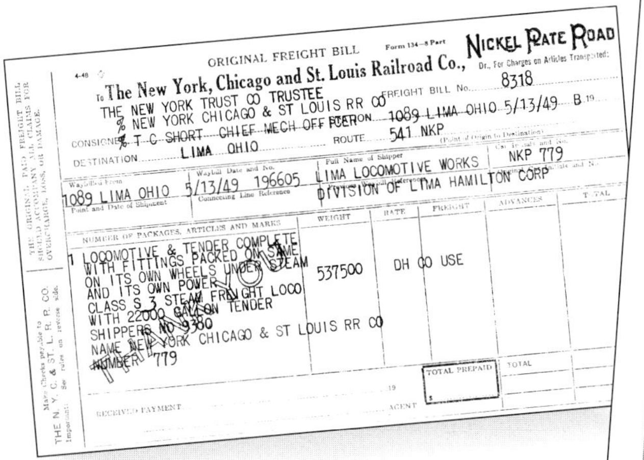

ORIGINAL FREIGHT BILL Form 134—8 Part NICKEL PLATE ROAD

4-48

To The New York, Chicago and St. Louis Railroad Co., Dr., For Charges on Articles Transported:

THE NEW YORK TRUST CO TRUSTEE
% NEW YORK CHICAGO & ST LOUIS RR CO
% T C SHORT CHIEF MECH OFFICER

FREIGHT BILL No. 8318

STATION 1089 LIMA OHIO 5/13/49 B 19

CONSIGNEE

ROUTE 541 NKP (Point of Origin to Destination)

DESTINATION LIMA OHIO

Car Initials and No. NKP 779

Full Name of Shipper LIMA LOCOMOTIVE WORKS DIVISION OF LIMA HAMILTON CORP

Waybilled From 1089 LIMA OHIO — Point and Date of Shipment

Waybill Date and No. 5/13/49 196605 — Connecting Line Reference

NUMBER OF PACKAGES, ARTICLES AND MARKS	WEIGHT	RATE	FREIGHT	ADVANCES	TOTAL
1 LOCOMOTIVE & TENDER COMPLETE WITH FITTINGS PACKED ON SAME ON ITS OWN WHEELS UNDER STEAM AND ITS OWN POWER CLASS S 3 STEAM FREIGHT LOCO WITH 22000 GALLON TENDER SHIPPERS NO 9380 NAME NEW YORK CHICAGO & ST LOUIS RR CO NUMBER 779	537500	DH CO USE			
			TOTAL PREPAID $	TOTAL	

19

RECEIVED PAYMENT ... AGENT

THE ORIGINAL PAID FREIGHT BILL SHOULD ACCOMPANY ALL CLAIMS FOR OVERCHARGE, LOSS, OR DAMAGE.

Make Checks payable to THE N. Y. C. & ST. L. R. R. CO.
Important: See rules on reverse side.

(above) NKP freight bill 8318 – dated May 13, 1949 – proved to be a significant slip of paper. It heralded the arrival of the Nickel Plate Road's final steam locomotive, and would serve as an inauspicious obituary for steam production at Lima.

(John B. Corns Collection)

(right) Lima-Hamilton indulged in what was perhaps wishful thinking less than a month after Nickel Plate 779 had been delivered, with this trade ad confirming the builder's stated belief in the future of steam propulsion. The message fell on ears deafened by the likes of the EMD 567 and Alco 244 diesel engines, and just over two years later locomotive production ceased in Lima.

(Author's Collection)

(below) When it was not performing at the Fair, NKP 779 was placed on static display at the Eastern Railroads' exhibit with the Nickel Plate's other equipment contributions (a PS-1 boxcar and an aluminum boxcar).

(John B. Corns Collection)

774

777
NICKEL PLATE ROAD

(NKP Photo by Willis A. McCaleb; John B. Corns Collection)

Radio Equipped

(left) NKP 774's cold water pump is drenching the ties as the S-3 leads a morning westbound train across the viaduct at Painesville, Ohio, in the summer of 1957. The ten S-3 Berkshires spent their nine-year careers on this picturesque stretch of Nickel Plate main line between Bellevue and Buffalo, venturing west to Fort Wayne on only rare occasions .

(below) The "Three Hooks" leads sister 779 and 86-car Buffalo-Chicago manifest NC-3 out of the siding at Perry, Ohio, on September 1, 1957, after a meet with eastbound passenger train number 6, the *City of Cleveland*. A look at the 777's freshly wiped cab numbers reveals where East End NKP crews got their inspiration for nicknaming this Berkshire.
(Both, John A. Rehor; John B. Corns Collection)

As a vital part of the Nickel Plate District's operations, 44 Berkshires were included in the Road's radio installation program after 1952 (26 on the Buffalo and Cleveland Divisions, and 18 on the Fort Wayne and Chicago Divisions). All except the 741 and 777-779 received Bendix sets (the four exceptions had Westinghouse gear). From the right angle, a radio-equipped Berkshire could be identified by the rectangular metal transmitter/ receiver/antenna box mounted on the engineer's side of the tender deck immediately behind the coal bunker slope sheet (see photo above). A pair of electrical conduits ran forward along the same side of the tender into the cab. Radio gear on a Berkshire also required a fourth 1000-watt turbogenerator atop the fireman's side of the firebox.

Table 7
NKP Berkshires Equipped with Radio

	Year Installed					
	1952	1953	1954	1955	1956	1957
715					●	
731					●	
733-734			●			
735-738		●				
740-742		●				
743	●					
744-746		●				
747						●
748		●				
749-750				●		
755-760				●		
761-762			●			
763		●				
764-766						●
767-768				●		
769	●					
770			●			
771-773		●				
774			●			
775-779		●				

(above) The 773 leads a westbound manifest across the Huron River on August 6, 1949, just 4-1/2 months into its nine-year career.

(Richard J. Cook, Sr.; Allen County Historical Society Collection)

In an industrial vista soon to be overwhelmed by a new highway viaduct, Berkshires 770 and 734 traverse Cleveland's Cuyahoga River valley with westbound manifest NC-3 on March 4, 1957.

(John A. Rehor; John B. Corns Collection)

(above) A westbound manifest behind S-3 number 774 overtakes Mikado 662 on local train 91 at Vermilion, Ohio, on September 14, 1957.

(John A. Rehor; John B. Corns Collection)

(left) The 779 takes on coal and water during a crew change at Chestnut Street in Conneaut, while en route to Buffalo with manifest KC-44.

(Jim Shaughnessy)

777

(facing page, top) The entire span of Lima's NKP Berkshire construction is captured fleetingly in this view at Vermilion, Ohio, as S-3 779 charges past 715, patriarch of the S-1 class.
(Philip T. Horning; John B. Corns Collection)

(facing page, bottom) The "Three Hooks" again, this time lifting a westbound extra out of the slight dip at Athol Springs, New York, at 12:20 pm on April 3, 1953.
(J. F. Beveridge; Paterson-George Collection)

(right) General Motors' second failed attempt to supplant NKP Berkshires rests at Hoopeston, Illinois, during May 1949 in the form of F7A demonstrators 800 and 801. While the messenger did not succeed, the inevitability of the message was clear. Within 18 months, black-and-gold GP7s would arrive on the property, and the Berkshires' hold on Nickel Plate manifests would begin to weaken.
(NKP Photo by John D. Burger; John B. Corns Collection)

steam locomotive's arrival in Chicago. The show was reported to be the greatest single vacation destination in the country in 1948, and proved so popular that an encore was staged beginning June 25, 1949, on the same 50-acre site. Between July 24 and August 2, NKP 779 was a featured performer in the theatrical "Wheels A-Rolling" pageant, one of the Fair's most popular attractions.

Even as the newest S-3 Berkshires ran their first revenue miles in May 1949, EMD was attempting to overcome the embarassment of the June 1948 F3 tests. Two blue and silver F7A demonstrators, numbered 800 and 801 as GM's affront to the 700-series Berkshires, were dispatched from La Grange to run a series of tests on the Nickel Plate's Peoria Division and Clover Leaf District during May. While NKP perceived the demonstrators' cab-unit carbodies as unsuitable for setoff and switching work, the machinery inside made an impression – especially in the context of replacing obsolete Mikes on the hilly Peoria and Michigan City-Indianapolis (IMC) Divisions.

When a baker's dozen 1500-hp EMD GP7 road switchers were ordered in 1950 as NKP's first road diesels, the S-3 Berkshires gained more grim cclcbrity – Lima's swan song had now become the Nickel Plate's last new steam power. Dieselization on the Nickel Plate was a selective, pragmatic process, though, and the Berkshires were allowed to thrive for the better part of the decade, earning their keep to the end just as Will Woodard, John Bernet, and the AMC had envisioned.

Roster of NKP S-3 Class Berkshires

Lima Locomotive Works, Construction Order 1209, March-May 1949.

Number	Builder's Number	Delivered	Disposition
770	9371	March 14, 1949	Retired 6/58. Sold for scrap 1/62
771	9372	March 18, 1949	Retired 4/58. Sold for scrap 1/62
772	9373	March 24, 1949	Retired 4/58. Sold for scrap 4/63
773	9374	March 25, 1949	Retired 4/58. Sold for scrap 4/63
774	9375	April 26, 1949	Retired 3/58. Sold for scrap 4/63
775	9376	May 4, 1949	Retired 3/58. Sold for scrap 12/63
776	9377	May 10, 1949	Retired 4/58. Sold for scrap 11/63
777	9378	May 9, 1949	Retired 6/58. Sold for scrap 4/61
778	9379	May 11, 1949	Retired 3/58. Sold for scrap 2/62
779	9380	May 13, 1949	Retired 3/58. Preserved at Lima, Ohio, 1963

Chapter 6

THE S-4 CLASS:

This 1949 portrait of Wheeling & Lake Erie Berkshire 6413, class K-1, illustrates many of the ways in which Wheeling 2-8-4s differed from their Nickel Plate cousins. Footboard pilot, Boxpok drivers, non-hinged main smokebox door, smaller sandbox, painted number plate, visorless headlight, and lack of feedwater heater were all quick spotting features. The fact that Timken was a large on-line W&LE customer was reflected in these locomotives, which were equipped with roller bearings from pilot truck to tender. Wheeling's use of Mars warning lights on steam locomotives would be reflected in NKP standard practice after 1950. The 6413 was regularly assigned to Wheeling directors' trains (and even the occasional excursion), and was therefore one of the first W&LE Berkshires to receive a Mars light before the NKP lease.
(Robert J. Foster; John B. Corns Collection)

For its part, Alco continued to build Super-Power Berkshires even after losing the Nickel Plate's remaining orders to Lima. Between 1937 and 1942, Schenectady delivered 32 Berks to the Wheeling & Lake Erie. Even though W&LE was within the NKP/C&O/Van Sweringen sphere of influence, its Berkshires remained distinctive machines. They shared the AMC's 69" drivers and 25x34" cylinders, but Wheeling played a political card by equipping its Berkshires with roller bearings all around, in deference to on-line shipper Timken. At a maximum total operational weight of 779,600 pounds, the heaviest Wheeling Berkshires were still 7,300 pounds lighter than an NKP S-1, and fully 29,300 pounds shy of an S-3.

WHEELING'S BERKS

The Wheeling Berkshires were built in four separate orders, although all were members of the road's K-1 class. Locomotives 6401-6410 arrived in April 1937 as Alco order S-1771. They were followed by 6411-6415 in December 1938 and January 1939 (order S-1804), and 6416-6422 in February and March 1941 (order S-1835). The final group of W&LE Berkshires arrived from Alco's Schenectady works in August and September 1942, numbered 6423-6432 as order S-1869.

All but five of the 32 locomotives shared the 64,100 pound tractive power of the NKP Berkshires. W&LE 801-803, 805, and 807 were rated at 65,400 pounds. These five had a boiler pressure of 250 psi and a 4.02 factor of adhesion, as compared to figures of 245 psi 4.10 for the other 27 K-1 Berkshires. This was due to their having been reboilered by Alco early in their careers, after potentially catastrophic metal fatigue was discovered around boiler rivet holes (this problem was not restricted to W&LE Alcos, as several other roads had contemporary locomotives reboilered to correct the situation). A Wheeling K-1 had a wheelbase just three inches longer than a Nickel Plate Berkshire's, and an overall length over coupler knuckles just under five inches longer. The tenders applied to the W&LE Berkshires (classed 22-RH by NKP) mirrored the Nickel Plate's adoption of a 22,000-gallon, 22-ton design.

Barely six months after brand-new S-3 Berkshires 770-779 arrived from Lima,

Nickel Plate took "delivery" of still more steam locomotives – the 151 engines conveyed on December 1, 1949, with NKP's lease of the Wheeling & Lake Erie. Included in that total were the 32 Alco Berkshires, which were assigned NKP class S-4. They retained their 6400-series Wheeling numbers but traded their tender initials for the Nickel Plate's imitation gold script. "NYC&STL" corporate initials appeared on the engines' sand domes and tenders, along with the "W&LE LESSOR" lettering common to all leased Wheeling equipment (or, in the case of 6416-6432, "W&LE SUB-LESSOR," since equipment trusts were still in effect on these locomotives). The 6400s were renumbered to NKP 801-832 in 1952.

(above) Down on the Wheeling's south end, K-1 number 6410 blasts from the "Rathole" at Warrenton, Ohio.
(Bob Redmond; John B. Corns Collection)

(below) W&LE Berkshire 6408 starts out of the yard at Brewster, Ohio, on the engine's maiden run in April 1937. The year before, Wheeling officials borrowed and tested Nickel Plate Alco Berkshire 704 prior to placing their own initial 2-8-4 order with the Schenectady builder.
(Glenn Grabill; John B. Corns Collection)

Even as Nickel Plate diesels (and some of the older Berkshires) made inroads on the Wheeling District, many of the ex-W&LE Berks remained in service without venturing beyond their old home rails. Only in 1957, when new EMD SD9s and Alco RSD12s arrived to take over the heavy mine runs that characterized the Wheeling District, did the S-4s face widespread retirement or, for five luckier classmembers, reassignment to the

W&LE 6408 takes a mid-day water break at Bellevue on September 2, 1946. Even from this angle, the Wheeling Berkshires were distinctive. Tender-top backup light contrasted with NKP's preference for rear bulkhead mount. Placing the locomotive's roadnumber on a removable metal plate facilitated tender swaps during shop work, without the need for a trip to the paint shop. All the shop crew had to do was replace the metal plate and two glass backup light inserts to transfer a tender from one Berkshire to another.

(Richard J. Cook, Sr.; Allen County Historical Society Collection)

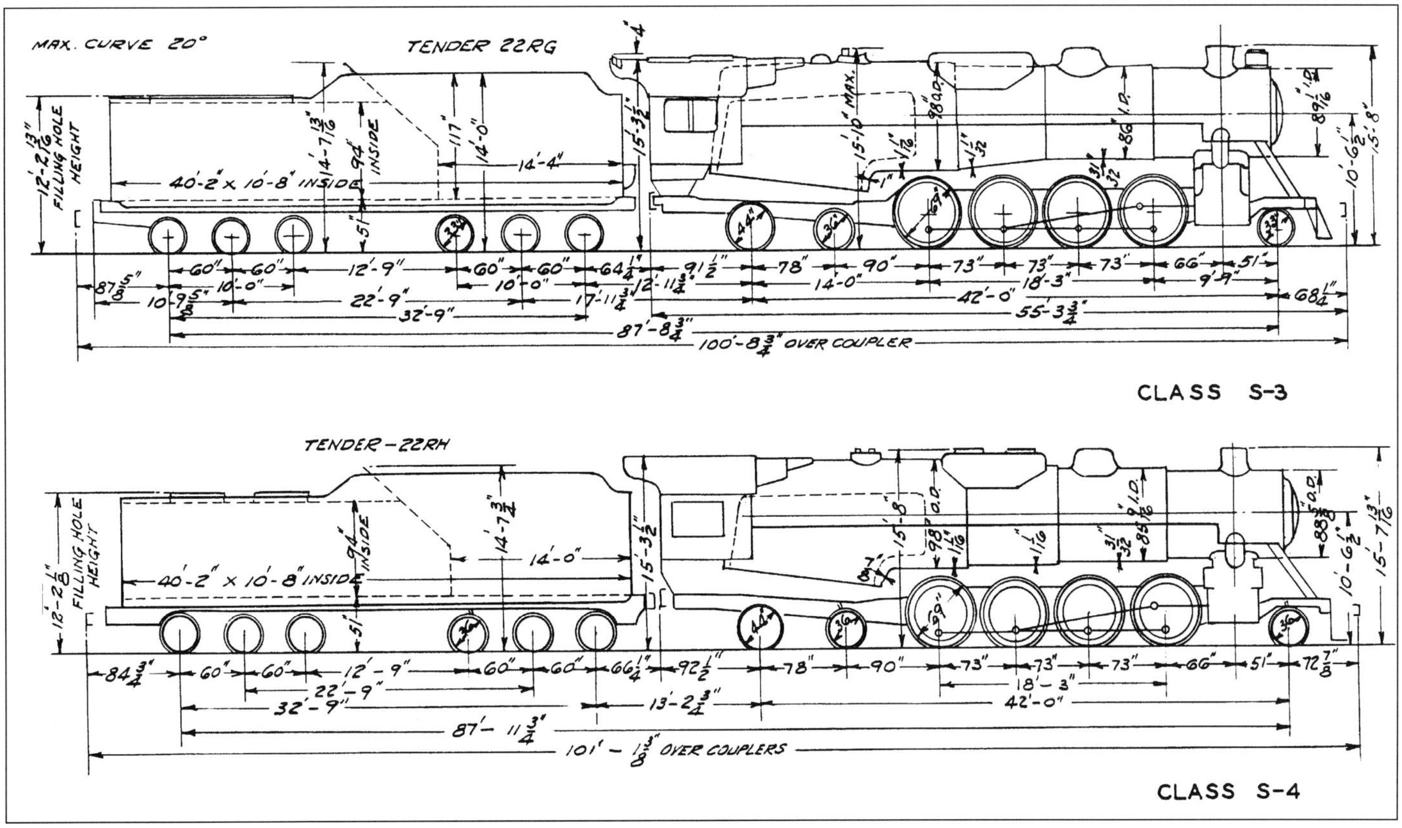

NKP diagrams offer a comparison of the S-3 and S-4 classes.

(Author's Collection)

(above) Even though it retains its original W&LE number, AMC-style "flying" numberboards and tender lettering mark the 6422 as a Nickel Plate locomotive in this view of eastbound train 84 at Yorkville, Ohio. This engine became NKP 822 in May 1952. Angled baffles behind footboards were another NKP modification to these ex-Wheeling locomotives, after 1950. Original pilots had flush steel plate behind footboards. Brewster shop memos show that new twin sealed beam headlights were chronically mis-aimed. *(J. J. Young, Jr.; John B. Corns Collection)*

(below) After they were renumbered into NKP's 800-series, the S-4 class still kept their rectangular number plates with painted yellow numerals – a practice inherited from former parent Wabash. Old W&LE numerals were simply painted over. Here, the former 6428 leads H-5b class 2-8-2 number 512 (Brooks, 1917) on 78-car eastbound train 86 at Harmon, Ohio, at 9:55 am on May 21, 1955. In six months' time, the venerable Mike would be renumbered 958 to clear a roster slot for subsequent GP9 orders. *(Fred Sankoff; The Sirman Collection)*

The Light From Mars

(NKP Photo by John D. Burger; John B. Corns Collection)

"Oscillating white headlights on engines so equipped will be displayed in addition to displaying of regular headlight when approaching and moving over highway crossings at grade except approaching passenger stations where stop is to be made."

From Special Instructions, Rule 17, NKP Ft. Wayne and Chicago Division Timetable, April 27, 1952.

For a company that rarely specified extra-cost options for its locomotives, the Nickel Plate became notable after 1947 as one of a very few railroads in the northeastern United States to make widespread use of oscillating warning lights. The devices made their Nickel Plate debut on the eleven Alco PA1s, delivered between December 1947 and March 1948 with model WR5000 fixtures bearing the trademark "The Light From Mars."

Around the same time that the Nickel Plate's "Bluebirds" were arriving, in early 1948, the Wheeling & Lake Erie was experimenting with oscillating light applications on some of its Berkshires. A portable barrel-shaped Pyle model 10350 light (designed for use on the rear of passenger trains) was installed, with a white lens, above the original headlight on the smokebox front of W&LE engine 6429. Wheeling 6413 (photo, at Brewster in April 1948) and 6428 are also known to have received white oscillating lights, but these were made by Mars in the same heavy duty housing that would become familiar to NKP observers a couple of years later. The prolonged vibration and exposed location of steam locomotive Mars light installations called for a specially insulated housing, designed to minimize shocks to the light's circuitry, gearing, and bulb filament that weren't present in diesel or passenger car applications. In the cases of W&LE 6413 and 6428, a single-light Mars mechanism was installed inside the cast aluminum housing, with a hinged clear outer lens 18" in diameter. Unlike the Pyle units, which used a rotating reflector to create a more-or-less circular gyrating light beam, Mars' patents enabled that company to market lights whose twin-cam geared motor linkages created a more effective (but more maintenance-intensive) oscillating figure-8 beam pattern.

Just as it had on the issue of diesel switcher safety cages, NKP followed the Wheeling's lead. Beginning with the installation of an oscillating-reflector Mars light on the 779 prior to its display at the 1949 Chicago Railroad Fair, Nickel Plate equipped every one of its own 80 Berkshires (along with the rest of the ex-Wheeling group following the December 1, 1949, W&LE lease) with a Mars SBWE-1000 warning light during 1950-51, following suit with many of the Mikes between 1952 and 1954, as well as five of the Hudsons (170-172, 174, and 176).

The model SBWE-1000 differed from the Mars lights applied to locomotives 6413, 6428, and 779 by using a 7-1/2" sealed beam bulb mounted in a cast metal frame, which was linked in turn to the patented oscillating mechanism.

In 1953, NKP considered replacing the SBWE-1000 Mars lights on its Berkshires with the dual beam units that had by then become the standard on the Road's GP7 road switchers. These vertically mounted Mars units oscillated two 7-1/2" sealed beams in unison. Drawings were prepared, but no conversions were made.

Installation of Mars lights on the Berkshires required the addition of a third turbogenerator to support the increased electrical load. The cast aluminum Mars housings were mounted on a steel shelf bolted to the smokebox door immediately above the original headlight, and remained part of the Berkshires' "profile" until the engines were withdrawn from service.

Westbound S-4 number 809 meets the 815 as it passes under the Baltimore & Ohio overpass at Lodi, Ohio, in 1957. *(Herbert H. Harwood, Jr.; John B. Corns Collection)*

"original" Nickel Plate between Bellevue and Buffalo (802, 805, 809, 810, and 814). The 814 even made it onto the cover of the employee-oriented January-February 1958 *Nickel Plate Road Magazine*, shown waiting to depart Buffalo's Abbott Road yard, just months before it made its last run on April 28, 1958.

As the other S-4 Berkshires were retired, some of their roller bearing-equipped Buckeye tender trucks were sent to Conneaut for placement under the tanks of 12 original NKP Berkshires. Between August 1957 and January 1958, the Road replaced the friction bearing Buckeyes on S-1 engines 718-720 and 725, S-2 743, and S-3 Berks 770-776. This "cannibalizing" followed a precedent set in June 1954, when the NKP began replacing the as-built screw-type power reverse gear on Lima S-1 Berkshires operating between Buffalo and Cleveland with Alco Type "G" hand-lever power reverse equipment. The Alco reverse gear was removed from ex-W&LE locomotives held for disposition, and was surely appreciated by Buffalo and Cleveland Division crews who found themselves assigned an S-1 for wayfreight duties.

The Nickel Plate regarded the ex-Wheeling Berkshires as more than just a parts source, however. On September 21, 1954, Brewster Shops General Foreman P. A. Ward issued a memo outlining a 36-point program of improvements and changes to the S-4 class. These ranged from upgrading cabs with side heaters, Insulmat insulation, and safety glass, to adding a generator and replacing the original trailing truck wheels with solid wheelsets. Stokers were converted to Hanna model H-4-1A, and Timken bearings were applied to the stoker worm shaft. In a corrosion reduction lesson stemming from a design change implemented on the S-3 class, the S-4s had their tender coping removed. This prevented rain and tank water from collecting atop the tender, where it invariably mixed with cinders, coal dust, and dirt to form a corrosive mud.

Steam operation on the Wheeling District expired in May 1957 as the final RSD12 and SD9 diesel road switchers were delivered. By May 9, the last S-4 in service at the Pine Valley terminal was 831. Later that day, the arrival of SD9 353 made the 1942 Alco redundant. It was retired in Brewster that June.

Roster of NKP S-4 Class Berkshires (former Wheeling & Lake Erie K-1 Class)

American Locomotive Company, Order S-1771, April 1937.

Final NKP Number	Builder's Number	Delivered	Disposition
801	68829	April 1937	Ex-6401. Retired 2/57. Sold for scrap 4/62
802	68830	April 1937	Ex-6402. Retired 3/58. To McLouth Steel 11/59
803	68831	April 1937	Ex-6403. Retired 12/56. Sold for scrap 5/61
804	68832	April 1937	Ex-6404. Retired 10/57. Sold for scrap 3/62
805	68833	April 1937	Ex-6405. Retired 5/58. Sold for scrap 10/61
806	68834	April 1937	Ex-6406. Retired 11/56. Sold for scrap 2/62
807	68835	April 1937	Ex-6407. Retired 11/56. Scrapped by NKP 6/61-4/64
808	68836	April 1937	Ex-6408. Retired 2/57. Sold for scrap 2/63
809	68837	April 1937	Ex-6409. Retired 3/58. Sold for scrap 10/61
810	68838	April 1937	Ex-6410. Retired 5/58. To McLouth Steel 11/59
American Locomotive Company, Order S-1804, December 1938-January 1939.			
811	69121	December 1938	Ex-6411. Retired 6/57. Sold for scrap 3/63
812	69122	December 1938	Ex-6412. Retired 10/57. Sold for scrap 5/61
813	69123	January 1939	Ex-6413. Retired 8/56. Sold for scrap 4/62
814	69124	January 1939	Ex-6414. Retired 5/58. Sold for scrap 12/63
815	69125	January 1939	Ex-6415. Retired 6/57. Sold for scrap 3/63
American Locomotive Company, Order S-1835, February-March 1941.			
816	69429	March 1941	Ex-6416. Retired 2/57. Sold for scrap 3/63
817	69430	March 1941	Ex-6417. Retired 8/57. Sold for scrap 9/61
818	69431	March 1941	Ex-6418. Retired 5/57. Sold for scrap 6/61
819	69432	March 1941	Ex-6419. Retired 2/57. Sold for scrap 5/61
820	69433	March 1941	Ex-6420. Retired 5/57. Scrapped by N&W 11/64
821	69434	March 1941	Ex-6421. Retired 9/56. Sold for scrap 5/61
822	69435	March 1941	Ex-6422. Retired 9/56. Sold for scrap 3/63
American Locomotive Company, Order S-1869, August-September 1942.			
823	70021	August 1942	Ex-6423. Retired 4/57. Sold for scrap 9/61
824	70022	August 1942	Ex-6424. Retired 4/57. Sold for scrap 10/61
825	70023	August 1942	Ex-6425. Retired 9/57. Sold for scrap 12/61
826	70024	August 1942	Ex-6426. Retired 10/56. Sold for scrap 2/63
827	70025	August 1942	Ex-6427. Retired 4/57. Sold for scrap 9/61
828	70026	August 1942	Ex-6428. Retired 3/57. Sold for scrap 1/62
829	70027	August 1942	Ex-6429. Retired 12/56. Scrapped by NKP 9/64
830	70028	August 1942	Ex-6430. Retired 4/57. Sold for scrap 9/61
831	70029	August 1942	Ex-6431. Retired 6/57. Sold for scrap 10/61
832	70030	September 1942	Ex-6432. Retired 2/57. Sold for scrap 1/62

(Alco Photo; John B. Corns Collection)

Chapter 7

MASTERS OF

Ninety-one Nickel Platers, most if not all former Wheeling & Lake Erie employees, gather at Brewster, Ohio, with S-4 Berkshires 811 and 822 and recently delivered EMD SW9 number 244 in this view circa 1953. That's a Pittsburgh & West Virginia F-M H20-44 switcher in the background, visiting via the Wheeling-P&WV connection at Pittsburgh Junction, Ohio.
(NKP Photo by Willis A. McCaleb; John B. Corns Collection)

The late J. J. Bernet's operating vision had certainly borne fruit, for by the time John W. Davin had assumed the Nickel Plate presidency in December 1942 and implemented sweeping postwar physical plant improvements, NKP was moving freight cars farther each day than any other Class 1 railroad in the east – 78.4 miles. Centralized Traffic Control (CTC) and Automatic Block Signaling (ABS) contributed to average freight train speeds in 1948 that were 20 percent faster than the national average. Eastbound freights (the direction of 63 percent of the Road's net ton-miles) moved at an average of 21.2 mph, while westbounds traveled at an average 18.7 mph. In the same period, average gross tonnage per train was 2,975 eastbound and 2,197

THEIR CRAFT

westbound. NKP's freight locomotives ran an average 164.4 miles per serviceable day in 1948, up from 132.1 in pre-depression 1927 and only 111.2 in 1937 (see Tables 8 and 9). The national average in 1948 was just below 92 miles per day for a freight locomotive.

The operational hub of the Nickel Plate was Bellevue, Ohio, where a 3,200-car yard classified all eastbound and westbound traffic. Reflecting the balance of traffic, NKP freight schedules favored eastbound trains which, on the Nickel Plate District, originated at Chicago's Calumet Yard (about 12 miles south of the Loop). Within the next 23 miles, trains picked up traffic from Chicago connections at Osborn, Indiana (via the Indiana Harbor Belt) and Hobart, Indiana (via Elgin, Joliet & Eastern), before proceeding the 115.6 miles to Fort Wayne's West Wayne yard for power change. With a freshly coaled and watered Berkshire, NKP trains proceeded 125 more miles east to Bellevue. Beyond Bellevue, the Nickel Plate's fast freights ran right through to Buffalo (245 miles) without changing power.

Frankfort, Indiana, served as the consolidating hub for Clover Leaf and Peoria Division freight trains, and crews at Frankfort regularly pre-blocked cars for forwarding at Buffalo (where NKP yard space was always at a premium). When the Wheeling & Lake Erie became part of the Nickel Plate in December 1949, it brought a markedly different traffic mix to the Road's operation, and gave the

Table 8
Selected NKP Operating Statistics, 1929-1947

	1929	1937	1946	1947
Revenue ton-miles	5,291,371,000	4,703,924,519	7,614,039,858	8,568,158,202
Net tons per freight train (average)	700	804	1,138	1,115
Gross ton-miles per freight train hour	26,279	37,693	48,782	48,696
Freight locomotive miles per day	132.1	111.2	150.7	164.4
Car-miles per freight car day	49.7	63.4	76.8	78.4
Freight train speed (mph, average)	14.2	18.1	18.9	19.2
Operating revenue	$56,385,456	$41,612,266	$74,332,966	$92,520,841
Operating expenses	$39,896,885	$28,395,481	$58,506,093	$67,245,584
Operating ratio	70.76	68.24	78.71	72.68
Net income	$7,390,042	$2,655,561	$5,567,790	$8,178,013

Sources: Railway Age, October 2, 1948; ICC Operating Statistics Of Large Steam Railways.

Table 9
NKP Postwar Freight Operations Financial Summary, 1946-1960

Year	Total Revenue Tons Carried	Merchandise Freight Revenue	Coal & Coke Freight Revenue	Net Income	Operating Ratio
1946	31,801,065	$63,584,766	$6,491,922	$5,567,790	78.71
1947	36,601,522	$79,313,745	$9,285,587	$8,178,013	72.68
1948	39,146,977	$96,600,355	$11,604,263	$15,627,252	67.78
1949†	47,883,263	$101,786,120	$20,009,271	$15,642,383	70.80
1950	51,693,416	$113,446,539	$26,698,410	$21,019,542	64.15
1951	55,755,656	$126,161,142	$26,950,080	$18,482,772	68.64
1952	51,634,103	$128,568,551	$26,374,961	$18,802,703	67.47
1953	52,001,892	$132,304,408	$27,141,585	$17,866,796	68.03
1954	43,106,307	$114,041,657	$24,123,664	$14,460,872	70.37
1955	52,012,074	$127,205,766	$27,943,469	$16,686,927	67.83
1956	54,383,218	$135,054,765	$31,648,635	$16,329,207	69.73
1957	49,006,786	$136,026,951	$29,057,745	$15,100,720	71.44
1958	37,772,332	$113,646,821	$20,664,160	$10,433,465	72.96
1959	41,182,080	$120,327,841	$22,507,883	$13,288,922	71.07
1960	42,103,871	$121,129,104	$21,035,623	$12,030,800	71.65

† Includes figures for W&LE from January 1 through November 30, 1949.

Source: NKP Annual Reports, 1946-1960.

NKP a long-sought source of originating business (prior to the Wheeling lease, three-quarters of the Nickel Plate's traffic was received from connections). Wheeling served the southeastern Ohio coal fields, and provided a Lake Erie outlet at Huron, Ohio, for both northbound coal and southbound iron ore traffic (the latter destined for steel mills in Cleveland and the Wheeling area, as well as in Pittsburgh via pool arrangements with the Pittsburgh & West Virginia).

In the Nickel Plate's environment of high locomotive utilization and high speed service, frequent and thorough maintenance was critically important. Backshops were maintained at Conneaut, Ohio, to handle virtually any work the Road's steam roster would need. Smaller shops were operated in Lima until 1956, and in Frankfort until the summer of 1957. The roundhouses at other terminals and division points performed running repairs and routine servicing. After the December 1, 1949, Wheeling lease (and until it was converted to diesel use in 1955), the sprawling 22-bay W&LE shops at Brewster, Ohio, augmented Conneaut as a major steam facility. Part of John Davin's postwar improvements were manifested in new roundhouses and engine terminals at Bellevue, Calumet, and

Bellevue, Ohio, was the Nickel Plate's operational hub, and the opening of the redeveloped yard and engine terminal in 1947 was a cornerstone of the Road's postwar transformation under president John W. Davin. S-3 number 774 (above) is readied for its next assignment with the facility's 500-ton coal dock and distinctive smokestack in the background. View from atop the coal dock (below), shows Bellevue's poured concrete and glass block roundhouse, 110-foot turntable, business car storage, and main-line tracks heading west.

(2 NKP Photos by Willis A. McCaleb; John B. Corns Collection)

Until Bellevue's new engine facilities were completed, Frankfort, Indiana, boasted the Nickel Plate's most modern roundhouse. With eleven of the 27 stalls as a backdrop, the last of the Alco-built Berkshires is lined up on the 100-foot table. Most of the roundhouse occupants in this 1945 view are Mikados, but K-1b Pacific 168 stands out with its red, white, and blue "Buy War Bonds" tender. Three NKP passenger locomotive tenders got this patriotic treatment during World War II, but the repainting program never included the Berkshires or Mikes.

(NKP Photo by William Rittase)

Fort Wayne, along with improvements to Conneaut shops.

The Berkshires were scheduled for a Class 3 overhaul every 200,000 miles – about once every two years at the Nickel Plate's level of power utilization, in which a 2-8-4 could easily rack up 8,000 or 9,000 miles a month.

During the six weeks it took the Nickel Plate's shop forces to perform a "Class 3" on one of the Berkshires, the locomotive was almost completely disassembled. The tender was disconnected from the engine and repaired separately, and then the boiler and frame were lifted away from the running gear. Perched on a pair of ungainly four-wheel shop trucks, what remained of the Berkshire was then moved into the boiler shop where it was stripped of cab, jacket, lagging, and appliances. All parts were chalked with the road number of their locomotive, and it was rare for components to be intermingled among NKP locomotives until the very last years of steam operations, when items were often scavenged from retired engines to keep the final few running. Wheels, axles, and frame were inspected for cracks; cylinders and frame alignment were checked; firebox staybolts and syphons were replaced; and grate and arch were inspected and repaired as necessary.

When a Berkshire emerged from Class 3 repairs, it was as close to being a new locomotive as possible. The skill and dedication of the Nickel Plate's shop forces – at Conneaut and throughout the system – ensured that the Berkshires' inherent abilities were optimized over the Road and reflected on the financial statement's bottom line.

168
714
DO NOT GET
ON OR OFF
TURNTABLE
WHILE MOVING

700
700

779
NKP
61660

Railroading was, and is, a 24-hour business. The almost ethereal quality of these splendid night portraits offers a marked contrast to the reality of servicing steam locomotives on hot, sultry northern Ohio nights.

(facing page) The oldest and youngest Nickel Plate Berkshires both passed through Conneaut's engine terminal on this July 1957 evening, with Alco 700 resting in the shadow of the main 500-ton coaling station. A separate coal tower was maintained adjacent to the main line at Chestnut Street. At 60 tons, it was sufficient to top up through-train locomotives (like the 779 on Buffalo-bound KC-44) without introducing the delay of a trip into the terminal proper.

(Both, Jim Shaughnessy)

(above) S-2 number 742 shows evidence of a recent cab numeral touch-up as the Berkshire simmers on Bellevue's ready track. As welcome as warm steam locomotive cabs could be during a Great Lakes winter, they offered no respite on humid summer nights.

(John A. Rehor; John B. Corns Collection)

The Nickel Plate followed a standard classification of repairs to locomotives and tenders. These ranged from Class 5 (tires turned or new; necessary repairs to boiler, machinery, and tender; one or more pairs of driver bearings refitted) to Class 1 (new boiler or back end; new or reset flues; tires new or turned; general repairs). Class 2 repairs meant a new firebox, or one or more boiler courses, along with flues and tires. A "Class 3" involved new or reset flues as well as firebox and boiler repairs and new tires. Costs for class repairs escalated rapidly through the 1950s, as parts became scarce. NKP even bought out the inventories of key hardware suppliers like New York Air Brake to assure availability of discontinued items. In 1954, Class 3 repairs cost $29,444 for an S-class Berk; $27,032 for an S-1; $28,642 for an S-2; $22,777 for a five-year-old S-3; and $32,912 for an S-4. Class 5 repair costs were $13,289; $13,882; $13,682; $15,349; and $13,033, respectively. The only Berkshires receiving Class 2 repairs in 1954 were the S-class, at a staggering $43,348 each. The need for Class 1 work was a death warrant for any NKP steam locomotive at that date.

Major shop work required a locomotive's boiler, firebox, and frame to be separated from its wheels. The boiler and firebox were mounted on temporary shop trucks, while the wheels went through the Wheel House, as in these views at Conneaut .

(left) With rods and brake rigging removed and sand and lube lines disconnected, the crane operator lifts the tenderless 759 off of its wheels.

(NKP Photo by John D. Burger; John B. Corns Collection)

(below left) Using an electrical process known as magnafluxing, two workers inspect a Berkshire's drivers for cracks and less obvious signs of metal fatigue in August 1946.

(below center) With magnafluxing and other maintenance completed, a drop table is used to lift one of the four pairs of drivers into position on overhauled S-2 number 746.

(below right) The drivers are eased into place.

(3 NKP Photos by George J. Morris; John B. Corns Collection)

(right) A set of drivers – in this instance, from a Mikado – are positioned on Frankfort's quartering and journal turning machine in 1943.

(NKP Photo by William Rittase; Jay Williams Collection)

NILES

While wheels and tender were being repaired in their own specialized shop areas, dozens of Nickel Platers – all specialists in their own right – overhauled the "heart" of the locomotive.

(left) The murky depths of Conneaut's backshop hold at least four Berkshires in various stages of heavy repair. The 763's boiler jacket has been reapplied, and the locomotive's air reservoirs appear ready for reinstallation. Whitewashed patches on frame assisted workers in locating stress fractures.

(below left) Using an inclined wooden plank as an impromptu scaffold, shop workers attend to an S-2's ashpan.
(2 NKP Photos by John D. Burger; John B. Corns Collection)

(below) S-2 number 763 poses on Conneaut's "new" (in 1945) four-track transfer table following Class 3 boiler, firebox, and frame work. Much remains to be done before the Berkshire can be returned to service, however. The Wheel House, Tender Shop, and Cab Shop must all make their contributions before final inspections and a trip through the Paint Shop can occur.
(NKP Photo by John D. Burger; John B. Corns Collection)

"Forty years on, when afar and asunder, parted are those who are singing today."

Edward Ernest Bowen, 1872
(From an English school song)

Chapter 8

TWILIGHT. . .

Conneaut, 1960. Spots of touch-up paint on cab and tender reveal that S-2 number 741 is, at least technically, one of 29 NKP Berkshires stored serviceable and still on the official roster. Dates stenciled on reservoir indicate that the 741 was shopped in April 1958, just before the ax fell on NKP mainline steam. According to NKP records, this engine received $29,380.26 worth of Class 5 repairs and was stored on April 8 without being broken in. After two years of storage, the locomotive's Mars light, builder's plate, and number plate have succumbed to "ivory hunters." The 741 met with a scrapper's torch at Cleveland in February 1962.
(John B. Corns Collection)

If, as *Trains Magazine* editor David P. Morgan wrote in 1955, the Nickel Plate's 2-8-4s "were engines a stockholder could admire on purely cash considerations," then the locomotives' esthetics were not lost on the legions of fans who, in the mid-1950s, realized that the Nickel Plate was staging the finale of high-speed steam railroading in the United States. Just as they had always done, the Berkshires led 100-car trains of perishables, piggyback, livestock, and merchandise – the "Million Dollar Manifests" – across the Nickel Plate at 60 miles per hour. That, in any event, was the official speed limit.

As black-and-gold diesel road switchers arrived, the Berkshires surrendered portions of their territory and began to be retired. S-1

AND SURVIVAL

number 730 was the last Berkshire to operate in NKP service over the St. Louis Division (the old Clover Leaf), when it led Second 98 from Charleston, Illinois, to Frankfort on July 18, 1955. Due to mechanical problems, it also had the distinction of being the first Berkshire officially retired, on August 29, 1955. Its fate sealed by the closure of Frankfort's steam backshops, the 730 was stored outdoors and gradually stripped until sold for scrap in November 1961. On September 6, 1956, S-1 number 720 towed sister engine 724 as the last active Berkshires through Lima, en route to reassignment on the Wheeling District (where they helped fill a void created by the W&LE's recently departed articulateds and ex-N&W Mountains). Berkshires 700 and 722 were still stored in Lima, though, and a similar eastbound deadhead move had occured one week earlier when S-1 engines 717 and 724 were ferried from Frankfort (see sidebar on page 106). GP9s had taken over from steam on the St. Louis Division, following the example of GP7s and RS3s that had combined to push the remaining NKP steam east of Fort Wayne or onto the yet-to-be-dieselized Wheeling & Lake Erie.

The last impediment to full freight dieselization on the Nickel Plate was the Chicago Division – 138.5 miles of which between West Wayne (Fort Wayne) and Calumet Yard was equipped with Automatic Train Stop (ATS). Use of this safety feature dated back to a 1922 I.C.C. order, and locomotives operating in this territory had to be outfitted with

Calling Ahead For Posterity

"On August 28, 1956, I was on the extra conductor's board at South Lima and was called to deadhead on passenger train number 9 to Frankfort, Indiana, the next morning to serve as conductor to accompany steam locomotives 717 and 724 to Lima.

"We departed Frankfort on the morning of Wednesday, August 29, with E. A. Donovan [at right in photo above, with conductor John H. Keller] as engineer on number 717, the lead engine. The fireman was Jack Gallant, and we also had a fireman on engine 724 to maintain heat in the boiler. When we arrived at Muncie, Indiana, I requested that Engineer Donovan stop just west of Kilgore Avenue. I then went to the phone at the crossing and talked with the dispatcher in Lima to see if there would be any train interference during the completion of our journey to Lima, 84 miles, and was assured that none was planned. I called Otto Etter, the telegraph operator at St. Marys, Ohio, and asked him to try and reach the photographer in that city and have him or her come down to the station and be prepared to take a series of photographs upon our arrival. At St. Marys station, Mr. Etter, a car inspector, and other employees had gathered so they would be a part of the photographs. We then proceeded to Lima without incident.

"There were two more steam locomotives brought from Frankfort to Lima in the same manner. I learned that they were on their way and was able to have a photographer from the *Lima News* greet them. It was always a joy when going on or off duty at South Lima to check the various types of steam locomotives in or near the roundhouse, and to look through the chain link fence which separated the Nickel Plate yard from Lima Locomotive Works and view the Shay locomotives switching within the Works."

John H. Keller

(Mr. Keller joined the Nickel Plate in July 1941, and his acute sense of history led to moments such as these being recorded. Aware that he was witnessing the end of an era, he worked diligently to ensure that significant reminders of the Nickel Plate's "Golden Era" were saved. The preservation of NKP S-3 number 779, caboose 1091, and business car 5 in Lima's Lincoln Park are perhaps the most tangible results of his efforts, and all students of the Nickel Plate are in his debt.)

ATS circuitry and an inductive pickup shoe. The only NKP diesels so equipped during the steam era were the eleven passenger-service PA1s, two NW2 switchers used in Chicago-area transfer service, and a pair of boiler-equipped GP9s held as protection for the four daily passenger trains. Freight trains over the Chicago Division still had to follow a Berkshire's drawbar.

That situation changed abruptly on June 13, 1958, when 30 blue-and-gold GP7s leased from Chesapeake & Ohio put their ATS gear to work on the Chicago Division – no doubt to EMD's tremendous satisfaction. In a March 17, 1959, letter to employees, NKP chairman Lynne White and president Felix Hales said the leased locomotives "made possible full dieselization of your Road at the level of business available throughout the balance of 1958. With receipt of ... 35 additional diesels in January and February 1959 ... the leased diesels have been returned to their owner. It is anticipated the 397 diesel units now owned by your Company will handle the volume of business foreseeable in the months ahead. Nevertheless, an adequate reserve of serviceable steam locomotives has been retained for use if required in the near future. Full dieselization contributed substantially to the favorable results of last year's operation."

The practice established by John Bernet and administered by Gus Ayers – in which fewer but more powerful locomotives handled steadily increasing traffic volumes – was continuing, but Bernet's Super-Power marvels had finally been supplanted by the diesel (see Table 10). John Bernet died in 1935, but had lived to see his S-class Berkshires begin their transformation of the Nickel Plate. John Davin passed away in 1949, just months after his S-3 Berks had been ordered. Gus Ayers retired in 1948, shortly after the S-3 order was placed. When he died in 1958, the Nickel Plate's motive power "old guard" was gone, and the pace of change accelerated.

By the time the leased C&O Geeps arrived, only 22 Berkshires remained in active service, and new NKP diesels had already taken over most assignments outside of the Chicago Division. Although steam had been responsible for 52.5% of Nickel Plate gross ton miles as late as March 1957, the end was in sight.

The "reclaiming" of steam locomotives was all in a day's work for scrappers like Luntz of Canton, Ohio, where these two poignant scenes were recorded in 1963. S-4 number 814 bleaches in the sun awaiting its turn under the torch, picked clean of almost every souvenir except – amazingly – one of its builder's plates. Amid piles of metal and hardware, the NKP Berk below is barely recognizable, having been reduced to little more than its running gear. *(Both, Gary Schmidt; John B. Corns Collection)*

The last ex-W&LE S-4 in service was the 810, which made its final Cleveland-Bellevue trip on May 8, 1958. S-2 number 740 led train 21 on June 18 as the final westbound steam movement from Bellevue. Two days later, the same engine made the final eastbound trip from Chicago's Calumet yard with train 32, arriving in Bellevue on June 21. The final Berkshire to run from Buffalo to Conneaut was the 719, on July 1. On July 2, 1958, S-2 number 746 became the last 2-8-4 to operate in Nickel Plate revenue service when it arrived in Conneaut from Bellevue. The distinction of being the final Berkshire in steam for NKP went to 765, which was used as a stationary boiler in the East Wayne roundhouse as late as December 5, 1958. It had also pulled the last steam-powered Nickel Plate passenger train: a June 11, 1958, East Wayne-Chicago excursion.

Table 10

NKP Locomotive Roster Summary, 1929-1959

Year	Qty Steam	Qty Diesel	Total
1929	465	–	465
1934	322	–	322
1937	304	–	304
1946	313	10	323
1947	296	41	337
1951	392 †	117	509
1953	329	152	481
1959	45 *	397	442

† Increase reflects lease of W&LE in 1949.
* Stored serviceable

Sources: NKP Comparisons of Locomotives – 1934, 1951, 1953, 1959; Railway Age, October 2, 1948.

In total, 45 steam locomotives, including 29 Berkshires, were stored serviceable between July 2, 1958, and August 31, 1960 (see Table 11). An extended economic downturn precluded any chance of reactivation before enough NKP diesels had entered service to meet reasonable traffic demands. Among these were ten GP18s assigned to the hallowed 700-series in June 1960. The stored Berkshires and their kin, along with the now-redundant coaling and water facilities, were finally written off as a $341,978 "loss on obsolete steam locomotive material due to dieselization."

Conspicuous in their absence from the stored-serviceable roster were any members of the S-3 class. Despite their relative youth, they were passed over for storage in favor of the 29 Berks with the lowest mileage since their most recent shopping. None of the stored Berks had run more than 100 miles. Fifteen had less than 30 post-service miles, including four (718, 741, 755, and 759) that had been placed in storage without any break-in mileage. Like 62 other Berkshires (and a total of 113 NKP steam locomotives), the S-3 class was relegated to dead storage (see Table 12). Except for a very fortunate few, this was merely a prelude to scrapping.

Table 11

NKP Berkshires Stored Serviceable On Official Roster July 2, 1958 - August 31, 1960

Qty	Class	Roadnumbers
8	S-1	718, 719, 720, 725, 727, 735, 737, 738.
21	S-2	740, 741, 742, 743, 745, 746, 747, 748, 749, 750, 751, 752, 754, 755, 756, 757, 759, 761, 762, 763, 765.

Source: NKP Comparison of Locomotives; Office of Chief Mechanical Officer; 1959.

Table 12

NKP Berkshires In Storage By Location, July 16, 1958

Location	Roadnumbers
Bellevue	713, 727, 740, 742, 744, 749, 750, 756, 757, 761, 762, 802, 810.
Brewster	702, 706, 708, 712, 725, 728, 729, 801, 803, 806-808, 811-813, 815-832.
Conneaut	700, 707, 709, 715, 718-720, 724, 726, 731-738, 741, 743, 746, 747, 751, 752, 754, 755, 758-760, 763, 770-779, 804, 805, 809, 814.
East Wayne	739, 745, 748, 753, 764-769.
Frankfort	711, 730.

Source: NKP "Recapitulation of Engines Stored" 7/16/58.

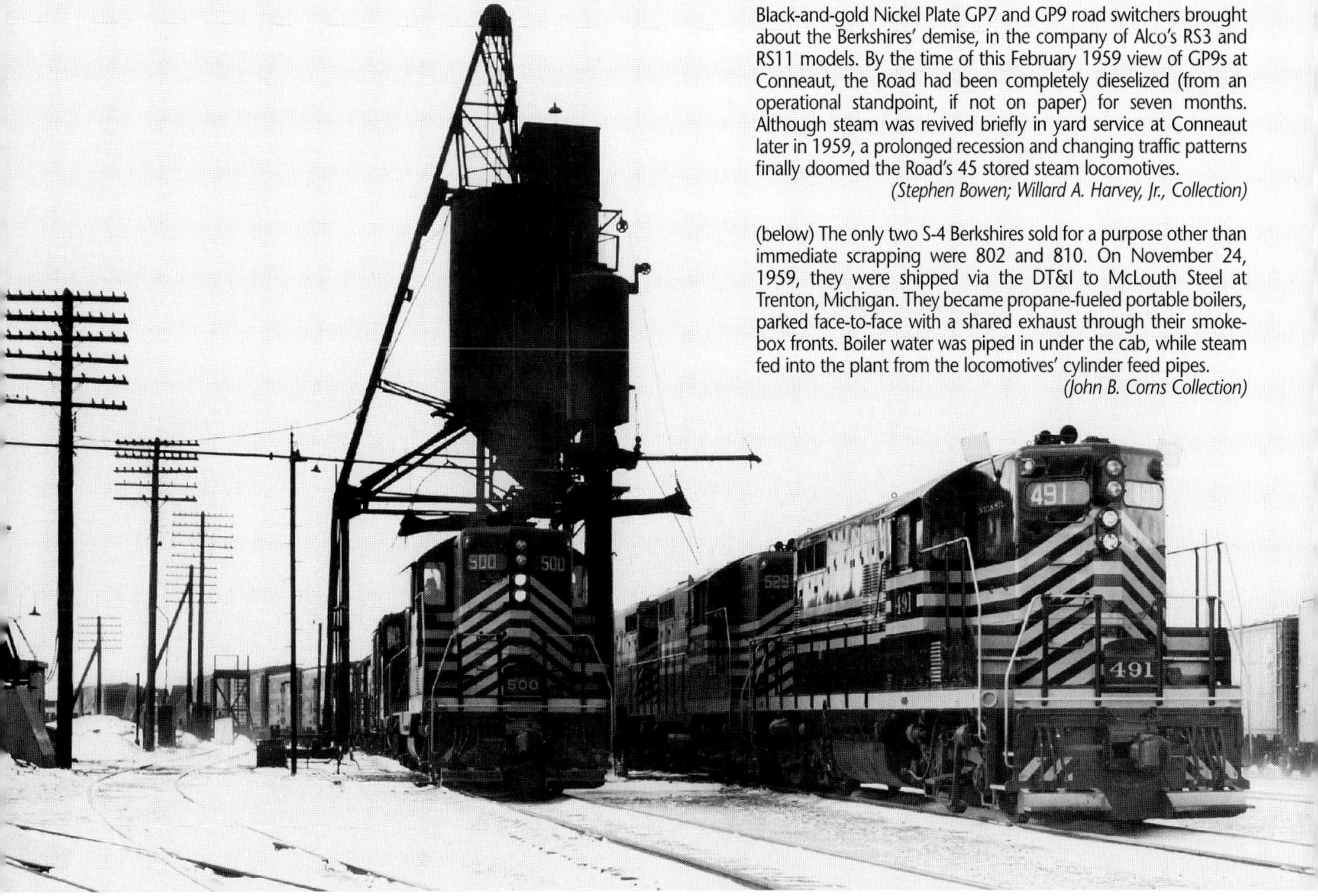

Black-and-gold Nickel Plate GP7 and GP9 road switchers brought about the Berkshires' demise, in the company of Alco's RS3 and RS11 models. By the time of this February 1959 view of GP9s at Conneaut, the Road had been completely dieselized (from an operational standpoint, if not on paper) for seven months. Although steam was revived briefly in yard service at Conneaut later in 1959, a prolonged recession and changing traffic patterns finally doomed the Road's 45 stored steam locomotives.

(Stephen Bowen; Willard A. Harvey, Jr., Collection)

(below) The only two S-4 Berkshires sold for a purpose other than immediate scrapping were 802 and 810. On November 24, 1959, they were shipped via the DT&I to McLouth Steel at Trenton, Michigan. They became propane-fueled portable boilers, parked face-to-face with a shared exhaust through their smokebox fronts. Boiler water was piped in under the cab, while steam fed into the plant from the locomotives' cylinder feed pipes.

(John B. Corns Collection)

Fresh from cosmetic refurbishing at Calumet shops, the 779 poses in the rain for her last official NKP portrait. Note missing builder's plate ... both were replaced with inaccurate reproductions provided by B-L-H. The gable-style numberplate is not original, but has been patched together using a plate from a 600-series H-6 Mikado, as evidenced by the "shadows" and extra holes left by removal of the original numerals.
(NKP Photo; John B. Corns Collection)

All told, six Lima-built Nickel Plate Berkshires escaped scrapping. Of these, three were enshrined in municipal parks and three were preserved in museums. Four of the surviving Berkshire sextet have led somewhat subdued retirements, but their mere preservation was victory enough.

In October 1964, NKP donated locomotive 755 to the city of Conneaut, Ohio, in fitting recognition of the role played by many of that city's residents in the maintenance and operation of the Road's Berkshires. The 755 became the most visible display of the Conneaut Railroad Historical Museum, parked across town on the grounds of the restored New York Central Station.

Like all of its Alco-built rostermates, the Nickel Plate's first 2-8-4 was unceremoniously scrapped. Fortunately, the Road's newest Berkshire fared somewhat better. NKP had retired the 779 at Conneaut in 1958, and there it would remain – exposed and vulnerable to "ivory hunters" – for almost five years. NKP executive vice-president Myron B. Phipps (who went on to become the Road's last president) donated the locomotive to the city of Lima in May 1963. After being moved from Conneaut and cosmetically refurbished by the Nickel Plate at Calumet shops in Chicago, the 779 was placed in protective storage at Lima's Ordnance Depot until a suitable display area could be prepared. By the summer of 1966, the city and the Allen County Historical Society (the Berkshire's custodians, primarily in the persons of John H. Keller and his son) had created a covered, fenced, floodlit display pavilion in Lima's Lincoln Park. On July 6, 1966, NKP SW8 number 112 pulled the Berkshire out of the Ordnance Depot and handed it over to Detroit, Toledo & Ironton NW2 number 912. The DT&I, through the efforts of vice-president of operations C. L. Towle, had installed a turnout and laid a short spur into the display area, and the NW2's accommodating crew ensured that the 2-8-4 was precisely spotted, with rods down, before they placed NKP caboose 1091 (Lafayette, 1882) and business car 5 (Pullman, 1883) behind it on the same spur. Lima's final steam locomotive had moved the last of its 677,095 miles, but it would figure prominently in the restorations of two older colleagues within a decade.

A handful of retired Berkshires lingered on former Nickel Plate property for a couple of years following the October 1964 N&W merger, and two of these were donated to off-line museums by the NKP's corporate successor. S-2 number 757 went to the Railroad Museum of Pennsylvania at Strasburg in 1966, and the 763 was shipped south on July 28, 1966, destined for display in N&W's hometown of Roanoke, Virginia. After cosmetic refurbishing at Norfolk & Western's Roanoke Shops, this favorite 2-8-4 of NKP crews was placed in the city's Wasena Park.

Roanoke's Berkshire returned north in 1974, when it was moved to the Central Railroad of New Jersey shops in Elizabethport, NJ, in anticipation of restoration to serve on the *American Freedom Train* (along with Conneaut's 755, which was never moved).

(above) The 763's refurbishing included "house number" numerals applied directly to the numberplate bracket, and a non-NKP visorless headlight with flat numberboards. Roanoke, June 1971. *(Howard W. Ameling)*

Fresh from a Class 3 with new roller-bearing tender trucks from an S-4, the 763 breaks in on Painesville trestle in September 1957. *(Don Wood)*

(above) A decade after the Nickel Plate dropped its fires, the 759's restoration crew takes a break at Conneaut in July 1968.
(John B. Corns Collection)

Agreements with both locomotives' owners had been reached, but plans for doubleheaded Berkshires to lead the *Freedom Train* were shelved before any restoration work had started. Clearance problems in Penn Central electrified territory and the costs and logistics of coaling a pair of 2-8-4s weighed against their restoration, as did political struggles within the project's management, so oil-burning Southern Pacific 4449 moved to the top of the list as power for the AFT (west of Chicago, anyway – ex-Reading T-1 number 2101 was rescued from a Baltimore scrapper and restored to power the AFT in the East). The 763 was sent back to Roanoke, where it remains today in the collection of the Virginia Museum of Transportation.

Befitting these locomotives' far-reaching role in the perfection of the Super-Power concept and the importance they played in the ascension of the Nickel Plate Road as a corporate dynamo, two of the surviving Berkshires were restored to operation.

The first of these was S-2 number 759, which underwent Class 3 repairs at Conneaut in early 1958. It emerged on May 20, just weeks before the C&O Geeps killed mainline NKP steam. Without the benefit of being broken in after shopping, the 759 became one of the 29 Berkshires listed on the roster as "stored serviceable" after July 1958. In early September 1962, fully two years after it had been "officially" retired, NKP sold 759 to steam preservationist and entrepreneur F. Nelson Blount. The locomotive was shipped to Blount's Monadnock, Steamtown & Northern site in North Walpole, New Hampshire, on October 16. This operation later became Steamtown USA, and moved across the Connecticut River to Bellows Falls, Vermont, but Blount's death in a plane crash on September 1, 1967, left an uncertain future for his locomotive collection. In 1966, businessman Ross Rowland, Jr., had created a successful steam excursion operation with his High Iron Company. In 1968, Rowland entered into an agreement with Steamtown whereby the High Iron Company would lease former NKP 759 and restore the Berkshire for operation on excursions throughout the East. The locomotive was promptly moved back to Conneaut, where Rowland's company had secured working space from Norfolk & Western in the old Nickel Plate shops. By the end of that summer, through the efforts of a crew that blended the talents of former Nickel Platers with eager and capable newcomers, 759 emerged from its old haunt ready for a second career as the first NKP steam locomotive to be resurrected since the end of steam and the disappearance of the Road itself on October 16, 1964. The 759 was formally christened in a ceremony at Conneaut on September 8, 1968, and a subsequent Conneaut-Buffalo break-in trip marked the first time in over a decade that a 2-8-4 had burnished Nickel Plate rails. The reborn Berkshire was soon leading sold-out fan trips to the likes of Roanoke, Horse Shoe Curve, Niagara Falls, and the Poconos.

The restoration of NKP 759 was a watershed in the then-fledgling steam preservation movement, as it was the first "big" locomotives brought back from extended retirement (as opposed to locomotives powering excursions on the Reading, CB&Q, GTW, Canadian National, and Union Pacific, among others, which had never really been retired). The talent assembled by Rowland to restore the 759, and the experience gained during the project, would figure prominently in his creation of the *American Freedom Train* of 1975-76, which in turn led to the restoration of two more "dead" steam locomotives – SP GS-4 4449, and Reading T-1 number 2101. The ripples of 759's rebirth would continue to spread, as big steam restorations continued – albeit infrequently – through the late 1990s.

The restored 759 was christened at Conneaut on September 8, 1968, and immediately went on the road for its new operators. On September 19, the High Iron Company - Steamtown Berkshire posed at Lebanon, New Jersey. Reproduction builder's plates were eventually applied.

(Don Wood)

(above) The *Golden Spike Centennial Limited's* 14 powder blue cars follow specially painted 759 and its High Iron Co. auxiliary tender westbound through Old Fort, Ohio, on May 5, 1969.
(Howard W. Ameling)

(below) On May 1, 1971, 759 led the last run of N&W's *Pocahontas* from Roanoke to Norfolk, Virginia. On April 24, the Berk led an N&W freight through Luray, Virginia, on its way south.
(Both, Thomas W. Dixon, Jr.)

The high point of 759's all-too-brief excursion career came when the locomotive led the May 1969 *Golden Spike Centennial Limited*, Rowland's round trip New York-Utah extravaganza operated to celebrate the driving of the Last Spike at Promontory on May 10, 1869. Artist Bob Lorenz's paint scheme for the train was reflected in powder blue panels and gold pinstriping applied to the Berkshire's tender, cab, cylinders, and domes, and "American Railroads 1869-1969" lettering on the tender. The 759 led the *Limited* out of Harmon, New York, on May 3, 1969, for a two-week out-and-back run to Promontory. (West of Kansas City, Union Pacific diesels and Northern 8444 pulled the train.) Several baggage/exhibit cars were included in the consist, and passengers had the option of paying $1,000 for the entire two-week "cruise" or $60 for an individual day's leg of the trip. The latter fare option proved more attractive, creating financial problems for the train which otherwise was an unequivocal public relations success. Logistical and operational experience gained from the *Golden Spike Centennial Limited* proved invaluable when the *American Freedom Train* debuted six years later.

The 759's other moment in the national spotlight came when the locomotive was leased by Norfolk & Western to lead the Roanoke-Norfolk leg of the final pre-Amtrak run of the railroad's Cincinnati-Norfolk *Pocahontas*, train number 4, on May 1, 1971 – it was even relettered N&W for the occasion. The Berkshire had blown its left cylinder head on February 20, 1971, as it led an excursion over N&W from Hagerstown to Roanoke, and continued mechanical and administrative troubles led to its eventual withdrawal from operation

The glory years ... you can bet that every one of the train's dutch doors is open as 759 assaults Horse Shoe Curve on the former Pennsylvania Railroad's Allegheny crossing on this September 13, 1970 excursion.

(John B. Corns)

At Bellevue on August 6, 1949, S-2 number 765 was just one of 80 Nickel Plate Berkshires helping the Road keep pace with post-war traffic. Celebrity was still three decades – and a change of identity – in the future.
(Richard J. Cook, Sr.; Allen County Historical Society Collection)

and a return to Steamtown in 1971. Following two brief reprieves (in 1974 and 1975), the locomotive was retired once again and placed on static display at Bellows Falls and, later, Scranton, Pennsylvania.

The second Nickel Plate Berkshire restored to operation was the 765, which took a more roundabout route to celebrity than 759. The city of Fort Wayne, Indiana, approached the Nickel Plate in 1962 and requested a Berkshire for civic display. Three 700s were still stored at East Wayne yard at the time (765, 766, and 767), and the city set its sights on the 767, since that engine had offically opened the Nickel Plate's 1955 grade separation through downtown Fort Wayne and was therefore seen to have played a part in the city's progress. (Interestingly, it was 765 that actually led the first "unofficial" run, a west-bound coal train, across the just-completed elevated route.) Unfortunately, the Nickel Plate's assessment of 767 revealed a deteriorating locomotive not suitable for much beyond scrap. It had suffered frame damage in a 1957 collision, and four years of outside storage had corroded the boiler jacket beyond mere cosmetic repair (the 767 lived on after a fashion, though, contributing components ranging from seats to water glasses to sister 759, which was being prepared for shipment to Steamtown at the time). The 765, on the other hand, had benefitted from two years' storage inside East Wayne roundhouse, and was in the best condition of the three extant Fort Wayne Berkshires. In order to appease the city, then, NKP quietly swapped numbers on Berkshires 765 and 767 in January 1963, and the "new" 767 (alias 765) was dedicated in its floodlit Lawton Park display site on May 4, 1963. The real 767 (renumbered 765) was sold for scrap that October.

Enter the Fort Wayne Railroad Historical Society, a non-profit group in possession of a new 25-year lease in which the Berkshire's municipal owner authorized a complete restoration for eventual excursion service. Masquerading as 767, Nickel Plate 765 remained on display until September 7, 1974, when temporary rails were laid across the city's Fourth Street and a front-end loader pulled the locomotive out of its chain-link enclosure and onto an adjacent Penn Central siding. With the locomotive returned to its correct number, the outdoor restoration proceeded as funds and volunteer labor permitted. By early 1978, a hydrostatic test had been made on the boiler. That September, the locomotive was fired for the first time in almost twenty years (it had been the last NKP Berkshire in steam, as a stationary boiler inside the East Wayne roundhouse on December 5, 1958). One year later, on September 1, 1979, Nickel Plate Berkshire 765 moved under its own power for the first time since June 14, 1958, albeit on just one quarter mile of track. The opportunity for an extended series of break-in trips came between May 4 and May 9, 1980, when Toledo, Peoria & Western president Robert McMillan authorized the Berkshire's use in regular TP&W freight and switching service. The locomotive performed just as Lima and the AMC had intended, and went on to become one of the benchmarks of "big steam" excursion operations through the 1980s and 1990s, traveling extensively through the East, South, and Midwest.

(above) Duck! Fort Wayne's new NKP passenger station and elevated right of way are officially opened in October 1955 as 767 drifts through a ceremonial ribbon, much to the dismay of at least one invited guest.

(John B. Corns Collection)

(below) The 765 spent eleven years in Fort Wayne's Lawton Park, disguised as the Berkshire that had earned a place in the city's history. Note the non-standard illuminated numberboard numerals.
(NKP Photo by Willis A. McCaleb; John B. Corns Collection)

756
Ted Rose '92

756
NICKEL PLATE ROAD

Epilogue

BERKSHIRES IN COLOR

(Preceding pages) In *"Hotshot"*, Ted Rose captures S-2 number 756 striding westward from Fostoria, Ohio, on a crisp winter morning in 1956. Squeaking snow, barking exhaust, throaty whistle, and whining turbogenerators all surge past our frigid vantage point ... then little more than the rythmic union of wheels and rail joints and a friendly wave from a Nickel Plater 100 cars back. That was Berkshire Country. Thanks, Ted.

With a low smoke plume and flapping number indicator attesting to the train's speed, S-2 number 745 whips through Knox, Indiana, with an eastbound manifest in May 1958.
(Jim Hawk; Howard W. Ameling Collection)

(above) S-2 number 758 heads east through South Gary, Indiana, on April 2, 1955.

(below) In its last winter of service, the 743 pauses at Conneaut in January 1958, in the company of an ex-Wheeling & Lake Erie S-4. *(Both, Louis A. Marre Collection)*

754
754

747
747
NKP
63504
NICKEL PLATE ROAD

(facing page, top) The 754 heads through Fostoria, Ohio, in May 1958. This engine was one of the 29 Berkshires stored serviceable between July 2, 1958, and August 31, 1960. It was scrapped by the NKP in October 1960.

(facing page, bottom) The Big Four (NYC) diamond at Claypool, Indiana, is about to feel the weight of S-2 number 747 and its westbound manifest in this April 1958 view on the Chicago Division. The Berkshire was retired three months later.

(above) Even the S-4s hauled piggyback on occasion. The 802 crosses the Huron River eastbound in the summer of 1957.

(left) S-2 number 746 eastbound at Kimball in 1957.

(All, Jim Hawk; Howard W. Ameling Collection)

(back cover, top) The 763 breaks in at Conneaut after a Class 3 overhaul in September 1957.

(TLC Collection)

(back cover, bottom) On May 16, 1969, the *Golden Spike Centennial Limited* returned east through Bellevue, Ohio, following the festivities honoring the 1869 driving of the last spike at Promontory, Utah.

(Dot Ameling)

Bibliography

Corns, John B.
Nickel Plate Road Publicity Photos, Volumes 1 and 2.
TLC Publishing Inc., 1996 and 1998.

Horning, Philip T., and Rehor, John A.
The Berkshire Era.
John A. Rehor, 1967.

Huddleston, Eugene L.
The Van Sweringen Berkshires.
N.J. International, Inc., 1986.

Lafayette Courier-Journal, Lafayette, Indiana.
"From Steam to Diesel – Era Ends."
August 20, 1955.

Lima Locomotive Works.
Delivery Records, 1922-1949.
Engineering Records, C.O. 1209, 1949.
Memorandum of Interview, July 7, 1948.

Lima News, Lima, Ohio.
"Diesels Put End To An Age."
September 18, 1955.

"Steam Era Ends On NKP."
September 6, 1956.

"Engine Gets Farewell Service Job."
February 1, 1957.

New York, Chicago & St. Louis Railroad.
Annual Reports, 1945-1961.

Comparisons of Locomotives.
Office of CMO, June 1934; February 1, 1951; February 12, 1953; March 1, 1959.

CMO Records and correspondence, 1934-1960

List of Officers, Agents, Stations, Etc., No. 5.
October 1951.

Locomotive Diagrams.
Mechanical Department, 1953 edition.

Timetable No. 140; April 27, 1952.
Ft. Wayne Division and Chicago Division.

Nickel Plate Road Historical & Technical Society, Inc.
Magazine. Winter 1978
Fall 1979
Spring 1980
Fall 1981
Spring 1983
Spring 1984

Reflections – Nickel Plate District
Second Printing, 1985.

Railway Age
"Engineering And Business Considerations Of The Steam Locomotive."
June 13, 1924.

"Performance Of Lima Locomotive No. 8000."
February 7, 1925.

"Lima Builds 2-8-4 Type Locomotive."
May 2, 1925.

"Naming The New Locomotive Types."
August 8, 1925.

"Tests Of 2-8-4 Locomotive On B&A."
September 12, 1925.

"Improved Operating Methods Accompany Improved Locomotives."
September 18, 1926.

"Western Railway Club Has Unusual Program."
December 25, 1926.

"A 4-8-4 Demonstrator Locomotive For Timken."
May 24, 1930.

"Timken Locomotive Completes First 100,000 Miles Of Service."
February 13, 1932.

"Nickel Plate Buys 2-8-4 Type Locomotives."
November 17, 1934.

"More Research Men Honored."
March 2, 1940.

"Tonnage With Speed On The Nickel Plate."
October 2, 1948.

Railway And Locomotive Engineering
"A 2-8-4 Locomotive For The Boston & Albany Railroad."
May 1925.

"Service Tests Of The Lima 2-8-4 Locomotive."
October 1925.

Railway Progress
"The Nickel Plate Will Take Its Berkshires."
December 1953.

Railway Review
"Boston & AlbanyTests A 2-8-4 Locomotive."
September 12, 1925.

Rehor, John A.
"The Engines That Saved A Railroad."
Trains Magazine, October 1962,
Kalmbach Publishing Company.

The Nickel Plate Story.
Kalmbach Publishing Company, 1965.

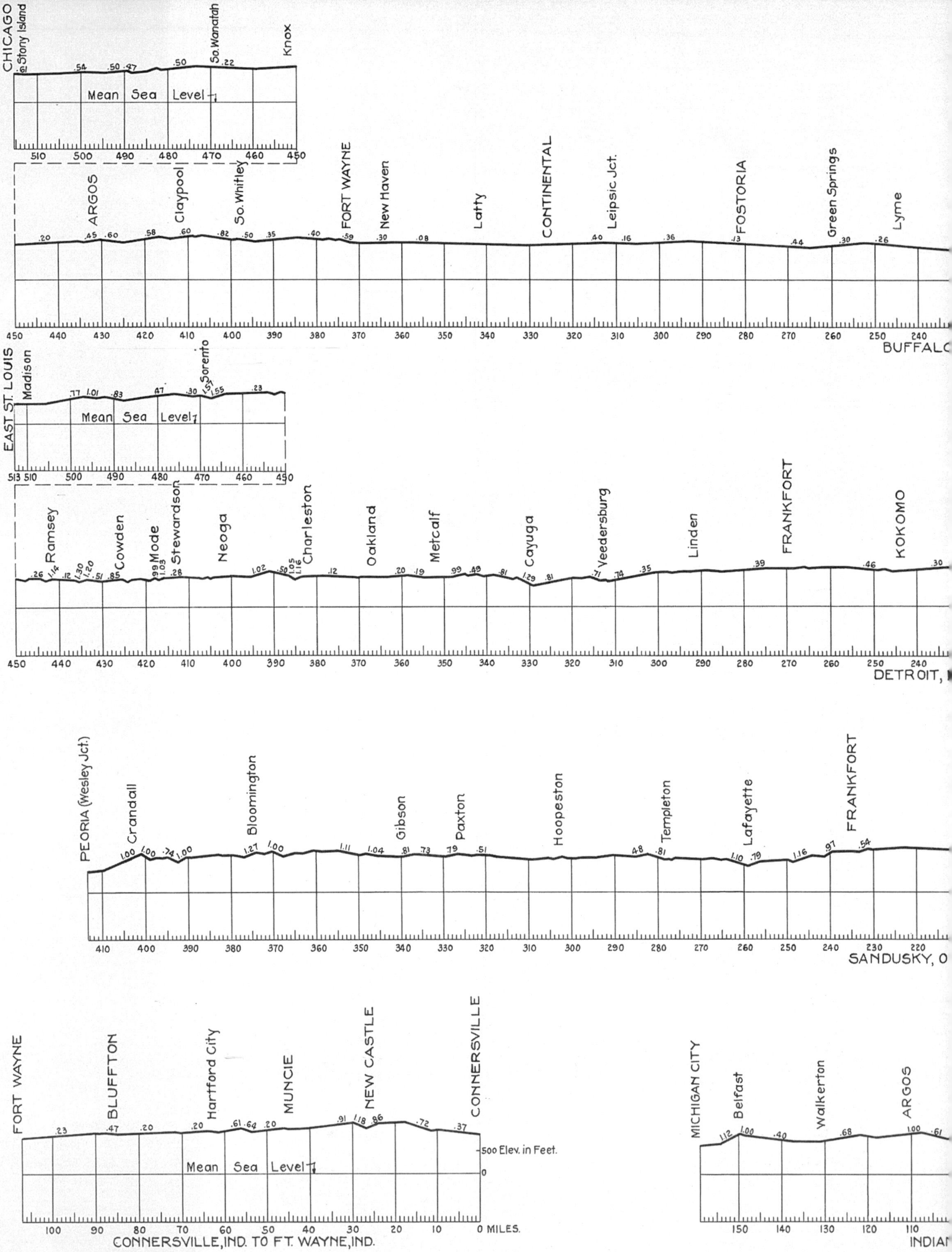
CHICAGO
Stony Island
So. Wanatah
Knox
Mean Sea Level
ARGOS
Claypool
So. Whitley
FORT WAYNE
New Haven
Latty
CONTINENTAL
Leipsic Jct.
FOSTORIA
Green Springs
Lyme
BUFFALO
EAST ST. LOUIS
Madison
Sorento
Mean Sea Level
Ramsey
Cowden
Mode
Stewardson
Neoga
Charleston
Oakland
Metcalf
Cayuga
Veedersburg
Linden
FRANKFORT
KOKOMO
DETROIT,
PEORIA (Wesley Jct.)
Crandall
Bloomington
Gibson
Paxton
Hoopeston
Templeton
Lafayette
FRANKFORT
SANDUSKY, O
FORT WAYNE
BLUFFTON
Hartford City
MUNCIE
NEW CASTLE
CONNERSVILLE
500 Elev. in Feet.
Mean Sea Level
0 MILES.
CONNERSVILLE, IND. TO FT. WAYNE, IND.
MICHIGAN CITY
Belfast
Walkerton
ARGOS
INDIAN